Human Molecular Genetics

Human Molecular Genetics

Ann Johnson

Larsen & Keller
www.larsen-keller.com

Human Molecular Genetics
Ann Johnson
ISBN: 978-1-64172-448-7 (Hardback)

⊟ Larsen & Keller

Published by Larsen and Keller Education,
5 Penn Plaza,
19th Floor,
New York, NY 10001, USA

Cataloging-in-Publication Data

Human molecular genetics / Ann Johnson.
      p. cm.
Includes bibliographical references and index.
ISBN 978-1-64172-448-7
1. Human molecular genetics. 2. Human genetics. 3. Molecular genetics. I. Johnson, Ann.
QH431 .H86 2020
611.018 16--dc23

For more information regarding Larsen and Keller Education and its products, please visit the publisher's website www.larsen-keller.com

Table of Contents

Preface

The biological field which deals with the study of inheritance in human beings is called human genetics. Human molecular genetics is one of its sub-fields where the functions and structure of human genes are studied at the molecular level. It uses tools from numerous fields such as genetics and molecular biology. Human molecular genetics finds application in the study of developmental biology as well as in the treatment of genetic diseases. There are a number of techniques which are used within this field such as forward genetics, reverse genetics and DNA replication. This book attempts to understand the multiple branches that fall under the discipline of human molecular genetics and how such concepts have practical applications. The topics covered herein deal with the core subjects of this field. This textbook will provide comprehensive knowledge to the readers.

A detailed account of the significant topics covered in this book is provided below:

Chapter 1- The field of biology which studies the function and structure of genes at a molecular level is known as molecular genetics. It is an amalgamation of the fields of molecular biology and genetics. This is an introductory chapter which will introduce briefly all these significant aspects of human molecular genetics.

Chapter 2- The structure of nucleic acids such as DNA and RNA is known as nucleic acid structure. Gene expression is the process that uses the information from a gene for synthesizing a functional gene product. This chapter closely examines the key concepts of nucleic acids and gene expression such as noncoding DNA, DNA replication and translation to provide an extensive understanding of the subject.

Chapter 3- The genome of Homo sapiens, which is made up of 23 pairs of chromosomes is known as the human genome. It consists of both protein coding DNA genes and non-coding DNA genes. A chromosome is a DNA molecule that contains a part or all of the genetic material of an organism. The topics elaborated in this chapter will help in gaining a better perspective about the human genome, the epigenome and the chromosomes.

Chapter 4- The direct interactions between cell surfaces are known as cell-cell interactions. They play an important role in the function and development of multicellular organisms. The defense system of the host which protects it from disease is known as the immune system. This chapter has been carefully written to provide an easy understanding of the varied facets of cell-cell interaction such as cell signalling and cell-cell adhesion, and biology of the immune system.

Chapter 5- There are many technologies which closely work with DNA such as polymerase chain reaction and DNA cloning. Polymerase chain reaction (PCR) is a method within the field of molecular biology where many copies of a particular DNA segment are made. The diverse DNA technologies such as nucleic acid hybridization and DNA repair have been thoroughly discussed in this chapter.

Chapter 6- In any particular population, the difference in DNA sequence among individuals is known as genetic variation. Mutation and genetic recombination are the two major sources of genetic variation. The chapter closely examines these key concepts of genetic variation to provide an extensive understanding of the subject.

It gives me an immense pleasure to thank our entire team for their efforts. Finally in the end, I would like to thank my family and colleagues who have been a great source of inspiration and support.

Ann Johnson

Introduction to Human Molecular Genetics

The field of biology which studies the function and structure of genes at a molecular level is known as molecular genetics. It is an amalgamation of the fields of molecular biology and genetics. This is an introductory chapter which will introduce briefly all these significant aspects of human molecular genetics.

Genetics

Genetics is the study of heredity in general and of genes in particular. Genetics forms one of the central pillars of biology and overlaps with many other areas, such as agriculture, medicine, and biotechnology.

Genetics arose out of the identification of genes, the fundamental units responsible for heredity. Genetics may be defined as the study of genes at all levels, including the ways in which they act in the cell and the ways in which they are transmitted from parents to offspring. Modern genetics focuses on the chemical substance that genes are made of, called deoxyribonucleic acid, or DNA, and the ways in which it affects the chemical reactions that constitute the living processes within the cell. Gene action depends on interaction with the environment. Green plants, for example, have genes containing the information necessary to synthesize the photosynthetic pigment chlorophyll that gives them their green colour. Chlorophyll is synthesized in an environment containing light because the gene for chlorophyll is expressed only when it interacts with light. If a plant is placed in a dark environment, chlorophyll synthesis stops because the gene is no longer expressed.

Genetics as a scientific discipline stemmed from the work of Gregor Mendel in the middle of the 19th century. Mendel suspected that traits were inherited as discrete units, and, although he knew nothing of the physical or chemical nature of genes at the time, his units became the basis for the development of the present understanding of heredity. All present research in genetics can be traced back to Mendel's discovery of the laws governing the inheritance of traits.

Methods in Genetics

Experimental Breeding

Genetically diverse lines of organisms can be crossed in such a way to produce different combinations of alleles in one line. For example, parental lines are crossed, producing an F1 generation, which is then allowed to undergo random mating to produce offspring that have purebreeding genotypes (i.e., AA, bb, cc, or DD). This type of experimental breeding is the origin of new plant and animal lines, which are an important part of making laboratory stocks for basic research. When applied to commerce, transgenic commercial lines produced experimentally are called genetically modified organisms (GMOs). Many of the plants and animals used by humans today (e.g., cows, pigs, chickens, sheep, wheat, corn (maize), potatoes, and rice) have been bred in this way.

Cytogenetic Techniques

Cytogenetics focuses on the microscopic examination of genetic components of the cell, including chromosomes, genes, and gene products. Older cytogenetic techniques involve placing cells in paraffin wax, slicing thin sections, and preparing them for microscopic study. The newer and faster squash technique involves squashing entire cells and studying their contents. Dyes that selectively stain various parts of the cell are used; the genes, for example, may be located by selectively staining the DNA of which they are composed. Radioactive and fluorescent tags are valuable in determining the location of various genes and gene products in the cell. Tissue-culture techniques may be used to grow cells before squashing; white blood cells can be grown from samples of human blood and studied with the squash technique. One major application of cytogenetics in humans is in diagnosing abnormal chromosomal complements such as Down syndrome (caused by an extra copy of chromosome 21) and Klinefelter syndrome (occurring in males with an extra X chromosome). Some diagnosis is prenatal, performed on cell samples from amniotic fluid or the placenta.

Biochemical Techniques

Biochemistry is carried out at the cellular or subcellular level, generally on cell extracts. Biochemical methods are applied to the main chemical compounds of genetics—notably DNA, RNA, and protein. Biochemical techniques are used to determine the activities of genes within cells and to analyze substrates and products of gene-controlled reactions. In one approach, cells are ground up and the substituent chemicals are fractionated for further analysis. Special techniques (e.g., chromatography and electrophoresis) are used to separate the components of proteins so that inherited differences in their structures can be revealed. For example, more than 100 different kinds of human hemoglobin molecules have been identified. Radioactively tagged compounds are valuable in studying the biochemistry of whole cells. For example, thymine is a compound found only in DNA; if radioactive thymine is placed in a tissue-culture medium in which cells are growing, genes use it to duplicate themselves. When cells containing radioactive thymine are analyzed, the results show that, during duplication, the DNA molecule splits in half, and each half synthesizes its missing components.

Chemical tests are used to distinguish certain inherited conditions of humans; e.g., urinalysis and blood analysis reveal the presence of certain inherited abnormalities—phenylketonuria (PKU), cystinuria, alkaptonuria, gout, and galactosemia. Genomics has provided a battery of diagnostic tests that can be carried out on an individual's DNA. Some of these tests can be applied to fetuses in utero.

Physiological Techniques

Physiological techniques, directed at exploring functional properties or organisms, are also used in genetic investigations. In microorganisms, most genetic variations involve some important cell function. Some strains of one bacterium (Escherichia coli), for example, are able to synthesize the vitamin thiamin from simple compounds; others, which lack an enzyme necessary for this synthesis, cannot survive unless thiamin is already present. The two strains can be distinguished by placing them on a thiamin-free mixture: those that grow have the gene for the enzyme, those that fail to grow do not. The technique also is applied to human cells, since many inherited human abnormalities are caused by a faulty gene that fails to produce a vital enzyme; albinism, which results from an inability to produce the pigment melanin in the skin, hair, or iris of the eyes, is an example of an enzyme deficiency in man.

Molecular Techniques

Although overlapping with biochemical techniques, molecular genetics techniques are deeply involved with the direct study of DNA. This field has been revolutionized by the invention of recombinant DNA technology. The DNA of any gene of interest from a donor organism (such as a human) can be cut out of a chromosome and inserted into a vector to make recombinant DNA, which can then be amplified and manipulated, studied, or used to modify the genomes of other organisms by transgenesis. A fundamental step in recombinant DNA technology is amplification. This is carried out by inserting the recombinant DNA molecule into a bacterial cell, which replicates and produces many copies of the bacterial genome and the recombinant DNA molecule (constituting a DNA clone). A collection of large numbers of clones of recombinant donor DNA molecules is called a genomic library. Such libraries are the starting point for sequencing entire genomes such as the human genome. Today genomes can be scanned for small molecular variants called single nucleotide polymorphisms, or SNPs ("snips"), which act as chromosomal tags to associated specific regions of DNA that have a property of interest and may be involved in a human disease or disorder.

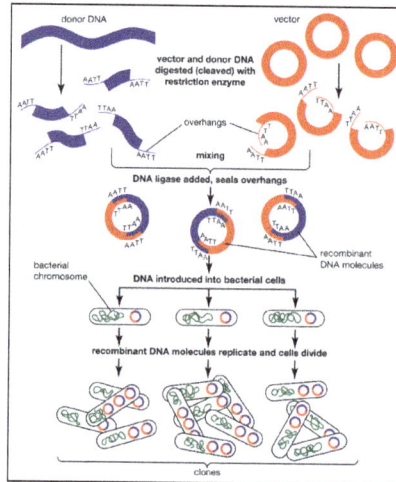

Steps involved in the engineering of a recombinant DNA molecule.

Immunological Techniques

Many substances (e.g., proteins) are antigenic; i.e., when introduced into a vertebrate body, they stimulate the production of specific proteins called antibodies. Various antigens exist in red blood cells, including those that make up the major blood groups of man (A, B, AB, O). These and other antigens are genetically determined; their study constitutes immunogenetics. Blood antigens of man include inherited variations, and the particular combination of antigens in an individual is almost as unique as fingerprints and has been used in such areas as paternity testing (although this approach has been largely supplanted by DNA-based techniques).

Immunological techniques are used in blood group determinations in blood transfusions, in organ transplants, and in determining Rhesus incompatibility in childbirth. Specific antigens of the human leukocyte antigen (HLA) genes are correlated with human diseases and disease predispositions. Antibodies also have a genetic basis, and their seemingly endless ability to match any antigen presented is based on special types of DNA shuffling processes between antibody genes. Immunology is also useful in identifying specific recombinant DNA clones that synthesize a specific protein of interest.

Mathematical Techniques

Because much of genetics is based on quantitative data, mathematical techniques are used extensively in genetics. The laws of probability are applicable to crossbreeding and are used to predict frequencies of specific genetic constitutions in offspring. Geneticists also use statistical methods to determine the significance of deviations from expected results in experimental analyses. In addition, population genetics is based largely on mathematical logic—for example, the Hardy-Weinberg equilibrium and its derivatives.

Bioinformatics uses computer-centred statistical techniques to handle and analyze the vast amounts of information accumulating from genome sequencing projects. The computer program scans the DNA looking for genes, determining their probable function based on other similar genes, and comparing different DNA molecules for evolutionary analysis. Bioinformatics has made possible the discipline of systems biology, treating and analyzing the genes and gene products of cells as a complete and integrated system.

Human Genetics

Human genetics is the study of the inheritance of characteristics by children from parents. Inheritance in humans does not differ in any fundamental way from that in other organisms.

The study of human heredity occupies a central position in genetics. Much of this interest stems from a basic desire to know who humans are and why they are as they are. At a more practical level, an understanding of human heredity is of critical importance in the prediction, diagnosis, and treatment of diseases that have a genetic component. The quest to determine the genetic basis of human health has given rise to the field of medical genetics. In general, medicine has given focus and purpose to human genetics, so the terms medical genetics and human genetics are often considered synonymous.

Fertilization, Sex Determination and Differentiation

A human individual arises through the union of two cells, an egg from the mother and a sperm from the father. Human egg cells are barely visible to the naked eye. They are shed, usually one at a time, from the ovary into the oviducts (fallopian tubes), through which they pass into the uterus. Fertilization, the penetration of an egg by a sperm, occurs in the oviducts. This is the main event of sexual reproduction and determines the genetic constitution of the new individual.

Human sex determination is a genetic process that depends basically on the presence of the Y chromosome in the fertilized egg. This chromosome stimulates a change in the undifferentiated gonad into that of the male (a testicle). The gonadal action of the Y chromosome is mediated by a gene located near the centromere; this gene codes for the production of a cell surface molecule called the H-Y antigen. Further development of the anatomic structures, both internal and external, that are associated with maleness is controlled by hormones produced by the testicle. The sex of an individual can be thought of in three different contexts: chromosomal sex, gonadal sex, and anatomic sex. Discrepancies between these, especially the latter two, result in the development of individuals with ambiguous sex, often called hermaphrodites. The phenomenon of homosexuality is of uncertain cause and is unrelated to the above sex-determining factors. It is of interest that in the absence of a male gonad (testicle) the internal and external sex anatomy is always female, even in the absence of a female ovary. A female without ovaries will, of course, be infertile and will not experience any of the female developmental changes normally associated with puberty. Such a female will often have Turner's syndrome.

If X-containing and Y-containing sperm are produced in equal numbers, then according to simple chance one would expect the sex ratio at conception (fertilization) to be half boys and half girls, or 1 : 1. Direct observation of sex ratios among newly fertilized human eggs is not yet feasible, and sex-ratio data are usually collected at the time of birth. In almost all human populations of newborns, there is a slight excess of males; about 106 boys are born for every100 girls. Throughout life, however, there is a slightly greater mortality of males; this slowly alters the sex ratio until, beyond the age of about 50 years, there is an excess of females. Studies indicate that male embryos suffer a relatively greater degree of prenatal mortality, so the sex ratio at conception might be expected to favour males even more than the 106 : 100 ratio observed at birth would suggest. Firm explanations for the apparent excess of male conceptions have not been established; it is possible that Y-containing sperm survive better within the female reproductive tract, or they may be a little more successful in reaching the egg in order to fertilize it. In any case, the sex differences are small, the statistical expectation for a boy (or girl) at any single birth still being close to one out of two.

During gestation—the period of nine months between fertilization and the birth of the infant—a remarkable series of developmental changes occur. Through the process of mitosis, the total number of cells changes from 1 (the fertilized egg) to about 2×10^{11}. In addition, these cells differentiate into hundreds of different types with specific func tions (liver cells, nerve cells, muscle cells, etc.). A multitude of regulatory processes, both genetically and environmentally controlled, accomplish this differentiation. Elucidation of the exquisite timing of these processes remains one of the great challenges of human biology.

Immunogenetics

Immunity is the ability of an individual to recognize the "self" molecules that make up one's own body and to distinguish them from such "nonself" molecules as those found in infectious microorganisms and toxins. This process has a prominent genetic component. Knowledge of the genetic and molecular basis of the mammalian immune system has increased in parallel with the explosive advances made in somatic cell and molecular genetics.

There are two major components of the immune system, both originating from the same precursor "stem" cells. The bursa component provides B lymphocytes, a class of white blood cells that, when appropriately stimulated, differentiate into plasma cells. These latter cells produce circulating soluble proteins called antibodies or immunoglobulins. Antibodies are produced in response to substances called antigens, most of which are foreign proteins or polysaccharides. An antibody molecule can recognize a specific antigen, combine with it, and initiate its destruction. This so-called humoral immunity is accomplished through a complicated series of interactions with other molecules and cells; some of these interactions are mediated by another group of lymphocytes, the T lymphocytes, which are derived from the thymus gland. Once a B lymphocyte has been exposed to a specific antigen, it "remembers" the contact so that future exposure will cause an accelerated and magnified immune reaction. This is a manifestation of what has been called immunological memory.

The thymus component of the immune system centres on the thymus-derived T lymphocytes. In addition to regulating the B cells in producing humoral immunity, the T cells also directly attack cells that display foreign antigens. This process, called cellular immunity, is of great importance in protecting the body against a variety of viruses as well as cancer cells. Cellular immunity is also the chief cause of the rejection of organ transplants. The T lymphocytes provide a complex network consisting of a series of helper cells (which are antigen-specific), amplifier cells, suppressor cells, and cytotoxic (killer) cells, all of which are important in immune regulation.

The Genetics of Antibody Formation

One of the central problems in understanding the genetics of the immune system has been in explaining the genetic regulation of antibody production. Immunobiologists have demonstrated that the system can produce well over one million specific antibodies, each corresponding to a particular antigen. It would be difficult to envisage that each antibody is encoded by a separate gene; such an arrangement would require a disproportionate share of the entire human genome. Recombinant DNA analysis has illuminated the mechanisms by which a limited number of immunoglobulin genes can encode this vast number of antibodies.

Each antibody molecule consists of several different polypeptide chains—the light chains (L) and the longer heavy chains (H). The latter determine to which of five different classes (IgM, IgG, IgA, IgD, or IgE) an immunoglobulin belongs. Both the L and H chains are unique among proteins in that they contain constant and variable parts. The constant parts have relatively identical amino acid sequences in any given antibody. The variable parts, on the other hand, have different amino acid sequences in each antibody molecule. It is the variable parts, then, that determine the specificity of the antibody.

Recombinant DNA studies of immunoglobulin genes in mice have revealed that the light-chain genes are encoded in four separate parts in germ-line DNA: A leader segment (L), a variable segment (V), a joining segment (J), and a constant segment (C). These segments are widely separated in the DNA of an embryonic cell, but in a mature B lymphocyte they are found in relative proximity (albeit separated by introns). The mouse has more than 200 light-chain variable region genes, only one of which will be incorporated into the proximal sequence that codes for the antibody production in a given B lymphocyte. Antibody diversity is greatly enhanced by this system, as the V and J segments rearrange and assort randomly in each B-lymphocyte precursor cell. The mechanisms by which this DNA rearrangement takes place are not clear, but transposons are undoubtedly involved. Similar combinatorial processes take place in the genes that code for the heavy chains; furthermore, both the light-chain and heavy-chain genes can undergo somatic mutations to create new antibody-coding sequences. The net effect of these combinatorial and mutational processes enables the coding of millions of specific antibody molecules from a limited number of genes. It should be stressed, however, that each B lymphocyte can produce only one antibody. It is the B lymphocyte population as a whole that produces the tremendous variety of antibodies in humans and other mammals.

Plasma cell tumours (myelomas) have made it possible to study individual antibodies, since these tumours, which are descendants of a single plasma cell, produce one antibody in abundance. Another method of obtaining large amounts of a specific antibody is by fusing a B lymphocyte with a rapidly growing cancer cell. The resultant hybrid cell, known as a hybridoma, multiplies rapidly in culture. Since the antibodies obtained from hybridomas are produced by clones derived from a single lymphocyte, they are called monoclonal antibodies.

The genetics of Cellular Immunity

As has been stated, cellular immunity is mediated by T lymphocytes that can recognize infected body cells, cancer cells, and the cells of a foreign transplant. The control of cellular immune reactions is provided by a linked group of genes, known as the major histocompatibility complex (MHC). These genes code for the major histocompatibility antigens, which are found on the surface of almost all nucleated somatic cells. The major histocompatibility antigens were first discovered on the leukocytes (white blood cells) and are therefore usually referred to as the HLA (human leukocyte group A) antigens.

The advent of the transplantation of human organs in the 1950s made the question of tissue compatibility between donor and recipient of vital importance, and it was in this context that the HLA antigens and the MHC were elucidated. Investigators found that the MHC resides on the short arm of chromosome 6, on four closely associated sites designated HLA-A, HLA-B, HLA-C, and HLA-D. Each locus is highly polymorphic; i.e., each is represented by a great many alleles within the human gene pool. These alleles, like those of the ABO blood group system, are expressed in codominant fashion. Because of the large number of alleles at each HLA locus, there is an extremely low probability of any two individuals (other than siblings) having identical HLA genotypes. (Since a person inherits one chromosome 6 from each parent, siblings have a 25 percent probability of having received the same paternal and maternal chromosomes 6 and thus of being HLA matched.)

Although HLA antigens are largely responsible for the rejection of organ transplants, it is obvious that the MHC did not evolve to prevent the transfer of organs from one person to another. Indeed, information obtained from the histocompatibility complex in the mouse (which is very similar in its genetic organization to that of the human) suggests that a primary function of the HLA antigens is to regulate the number of specific cytotoxic T killer cells, which have the ability to destroy virus-infected cells and cancer cells.

The Genetics of Human Blood

More is known about the genetics of the blood than about any other human tissue. One reason for this is that blood samples can be easily secured and subjected to biochemical analysis without harm or major discomfort to the person being tested. Perhaps a more cogent reason is that many chemical properties of human blood display relatively simple patterns of inheritance.

Blood Types

Certain chemical substances within the red blood cells (such as the ABO and MN substances) may serve as antigens. When cells that contain specific antigens are introduced into the body of an experimental animal such as a rabbit, the animal responds by producing antibodies in its own blood.

In addition to the ABO and MN systems, geneticists have identified about 14 blood-type gene systems associated with other chromosomal locations. The best known of these is the Rh system. The Rh antigens are of particular importance in human medicine. Curiously, however, their existence was discovered in monkeys. When blood from the rhesus monkey (hence the designation Rh) is injected into rabbits, the rabbits produce so-called Rh antibodies that will agglutinate not only the red blood cells of the monkey but the cells of a large proportion of human beings as well. Some people (Rh-negative individuals), however, lack the Rh antigen; the proportion of such persons varies from one human population to another. Akin to data concerning the ABO system, the evidence for Rh genes indicates that only a single chromosome locus (called r) is involved and is located on chromosome 1. At least 35 Rh alleles are known for the r location; basically the Rh-negative condition is recessive.

A medical problem may arise when a woman who is Rh-negative carries a fetus that is Rh-positive. The first such child may have no difficulty, but later similar pregnancies may produce severely anemic newborn infants. Exposure to the red blood cells of the first Rh-positive fetus appears to immunize the Rh-negative mother, that is, she develops antibodies that may produce permanent (sometimes fatal) brain damage in any subsequent Rh-positive fetus. Damage arises from the scarcity of oxygen reaching the fetal brain because of the severe destruction of red blood cells. Measures are available for avoiding the severe effects of Rh incompatibility by transfusions to the fetus within the uterus; however, genetic counselling before conception is helpful so that the mother can receive Rh immunoglobulin immediately after her first and any subsequent pregnancies involving an Rh-positive fetus. This immunoglobulin effectively destroys the fetal red blood cells before the mother's immune system is stimulated. The mother thus avoids becoming actively immunized against the Rh antigen and will not produce antibodies that could attack the red blood cells of a future Rh-positive fetus.

Serum Proteins

Human serum, the fluid portion of the blood that remains after clotting, contains various proteins that have been shown to be under genetic control. Study of genetic influences has flourished since the development of precise methods for separating and identifying serum proteins. These move at different rates under the impetus of an electrical field (electrophoresis), as do proteins from many other sources (e.g., muscle or nerve). Since the composition of a protein is specified by the structure of its corresponding gene, biochemical studies based on electrophoresis permit direct study of tissue substances that are only a metabolic step or two away from the genes themselves.

Electrophoretic studies have revealed that at least one-third of the human serum proteins occur in variant forms. Many of the serum proteins are polymorphic, occurring as two or more variants with a frequency of not less than 1 percent each in a population. Patterns of polymorphic serum protein variants have been used to determine whether twins are identical (as in assessing compatibility for organ transplants) or whether two individuals are related (as in resolving paternity suits). Whether the different forms have a selective advantage is not generally known.

Much attention in the genetics of substances in the blood has been centred on serum proteins called haptoglobins, transferrins (which transport iron), and gamma globulins (a number of which are known to immunize against infectious diseases). Haptoglobins appear to relate to two common alleles at a single chromosome locus; the mode of inheritance of the other two seems more complicated, about 18 kinds of transferrins having been described. Like blood-cell antigen genes, serum-protein genes are distributed worldwide in the human population in a way that permits their use in tracing the origin and migration of different groups of people.

Hemoglobin

Hundreds of variants of hemoglobin have been identified by electrophoresis, but relatively few are frequent enough to be called polymorphisms. Of the polymorphisms, the alleles for sickle-cell and thalassemia hemoglobins produce serious disease in homozygotes, whereas others (hemoglobins C, D, and E) do not. The sickle-cell polymorphism confers a selective advantage on the heterozygote living in a malarial environment; the thalassemia polymorphism provides a similar advantage.

Influence of the Environment

Gene expression occurs only after modification by the environment. A good example is the recessively inherited disease called galactosemia, in which the enzyme necessary for the metabolism of galactose—a component of milk sugar—is defective. The sole source of galactose in the infant's diet is milk, which in this instance is toxic. The treatment of this most serious disease in the neonate is to remove all natural forms of milk from the diet (environmental manipulation) and to substitute a synthetic milk lacking galactose. The infant will then develop normally but will never be able to tolerate foods containing lactose. If milk was not a major part of the infant's diet, however, the mutant gene would never be able to express itself, and galactosemia would be unknown.

Another way of saying this is that no trait can exist or become actual without an environmental contribution. Thus, the old question of which is more important, heredity or environment, is without meaning. Both nature (heredity) and nurture (environment) are always important for every human attribute.

But this is not to say that the separate contributions of heredity and environment are equivalent for each characteristic. Dark pigmentation of the iris of the eye, for example, is under hereditary control in that one or more genes specify the synthesis and deposition in the iris of the pigment (melanin). This is one characteristic that is relatively independent of such environmental factors as diet or climate; thus, individual differences in eye colour tend to be largely attributable to hereditary factors rather than to ordinary environmental change.

On the other hand, it is unwarranted to assume that other traits (such as height, weight, or intelligence) are as little affected by environment as is eye colour. It is very easy to gather information that tall parents tend, on the average, to have tall children (and that short parents tend to produce short children), properly indicating a hereditary contribution to height. Nevertheless, it is equally manifest that growth can be stunted in the environmental absence of adequate nutrition. The dilemma arises that only the combined, final result of this nature-nurture interaction can be directly observed. There is no accurate way (in the case of a single individual) to gauge the separate contributions of heredity and environment to such a characteristic as height. An inferential way out of this dilemma is provided by studies of twins.

Fraternal Twins

Usually a fertile human female produces a single egg about once a month. Should fertilization occur (a zygote is formed), growth of the individual child normally proceeds after the fertilized egg has become implanted in the wall of the uterus (womb). In the unusual circumstance that two unfertilized eggs are simultaneously released by the ovaries, each egg may be fertilized by a different sperm cell at about the same time, become implanted, and grow, to result in the birth of twins.

Twins formed from separate eggs and different sperm cells can be of the same or of either sex. No matter what their sex, they are designated as fraternal twins. This terminology is used to emphasize that fraternal twins are genetically no more alike than are siblings (brothers or sisters) born years apart. Basically they differ from ordinary siblings only in having grown side by side in the womb and in having been born at approximately the same time.

Identical Twins

In a major nonfraternal type of twinning, only one egg is fertilized, but during the cleavage of this single zygote into two cells, the resulting pair somehow become separated. Each of the two cells may implant in the uterus separately and grow into a complete, whole individual. In laboratory studies with the zygotes of many animal species, it has been found that in the two-cell stage (and later) a portion of the embryo, if separated under the microscope by the experimenter, may develop into a perfect, whole individual. Such splitting occurs spontaneously at the four-cell stage in some organisms (e.g., the armadillo) and has been accomplished experimentally with the embryos of salamanders, among others.

The net result of splitting at an early embryonic stage may be to produce so-called identical twins. Since such twins derive from the same fertilized egg, the hereditary material from which they originate is absolutely identical in every way, down to the last gene locus. While developmental and genetic differences between one "identical" twin and another still may arise through a number of processes (e.g., mutation), these twins are always found to be of the same sex. They are often breathtakingly similar in appearance, frequently down to very fine anatomic and biochemical details (although their fingerprints are differentiable).

Diagnosis of Twin Types

Since the initial event in the mother's body (either splitting of a single egg or two separate fertilizations) is not observed directly, inferential means are employed for diagnosing a set of twins as fraternal or identical. The birth of fraternal twins is frequently characterized by the passage of two separate afterbirths. In many instances, identical twins are followed by only a single afterbirth, but exceptions to this phenomenon are so common that this is not a reliable method of diagnosis.

The most trustworthy method for inferring twin type is based on the determination of genetic similarity. By selecting those traits that display the least variation attributable to environmental influences (such as eye colour and blood types), it is feasible, if enough separate chromosome loci are considered, to make the diagnosis of twin type with high confidence. HLA antigens, which, as stated above, are very polymorphic, have become most useful in this regard.

Inferences from Twin Studies

Metric (Quantitative) Traits

By measuring the heights of a large number of ordinary siblings (brothers and sisters) and of twin pairs, it may be shown that the average difference between identical twins is less than half the difference for all other siblings. Any average differences between groups of identical twins are attributable with considerable confidence to the environment. Thus, since the sample of identical twins who were reared apart (in different homes) differed little in height from identicals who were raised together, it appears that environmental-genetic influences on that trait tended to be similar for both groups.

Yet, the data for like-sexed fraternal twins reveal a much greater average difference in height (about the same as that found between ordinary siblings reared in the same home at different ages). Apparently the fraternal twins were more dissimilar than identicals (even though reared together) because the fraternals differed more from each other in genotype. This emphasizes the great genetic similarity between identicals. Such studies can be particularly enlightening when the effects of individual genes are obscured or distorted by the influence of environmental factors on quantitative (measurable) traits (e.g., height, weight, and intelligence).

Any trait that can be objectively measured in identical and fraternal twins can be scrutinized for the particular combination of hereditary and environmental influences that impinge upon it. The effect of environment on identical twins reared apart is suggested by their relatively great average difference in body weight as compared with identical twins reared together. Weight appears to be more strongly modified by environmental variables than is height.

Study of comparable characteristics among farm animals and plants suggests that such quantitative human traits as height and weight are affected by allelic differences at a number of chromosome locations—that they are not simply affected by genes at a single locus. Investigation of these gene systems with multiple locations (polygenic systems) is carried out largely through selective-breeding experiments among large groups of plants and lower animals. Human beings select their mates in a much freer fashion, of course, and polygenic studies among people are thus severely limited.

Intelligence is a very complex human trait, the genetics of which has been a subject of controversy for some time. Much of the controversy arises from the fact that intelligence is so difficult to define. Information has been based almost entirely on scores on standardized IQ tests constructed by psychologists; in general, such tests do not take into account cultural, environmental, and educational differences. As a result, the working definition of intelligence has been "the general factor common to a large number of diverse cognitive (IQ) tests." Even roughly measured as IQ, intelligence shows a strong contribution from the environment. Fraternal twins, however, show relatively great dissimilarity in IQ, suggesting an important contribution from heredity as well. In fact, it has been estimated that, on the average, between 60 and 80 percent of the variance in IQ test scores could be genetic. It is important to note that intelligence is polygenically inherited and that it has the highest degree of assortative mating of any trait; in other words, people tend to mate with people having similar IQs. Moreover, twin studies involving psychological traits should be viewed with caution; for example, since identical twins tend to be singled out for special attention, their environment should not be considered equivalent even to that of other children raised in their own family.

Since the time of Galton, generalizations have been repeatedly made about racial differences in intelligence, with claims of genetic superiority of some races over others. These generalizations fail to recognize that races are composed of individuals, each of whom has a unique genotype made up by genes shared with other humans, and that the sources of intraracial variation are more numerous than those producing interracial differences.

Other Traits

For traits of a more qualitative (all-or-none) nature, the twin method can also be used in efforts to assess the degree of hereditary contribution. Such investigations are based on an examination of cases in which at least one member of the twin pair shows the trait. It was found in one study, for example, that in about 80 percent of all identical twin pairs in which one twin shows symptoms of the psychiatric disorder called schizophrenia, the other member of the pair also shows the symptoms; that is, the two are concordant for the schizophrenic trait. In the remaining 20 percent, the twins are discordant; that is, one lacks the trait. Since identical twins often have similar environments, this information by itself does not distinguish between the effects of heredity and environment. When pairs of like-sexed fraternal twins reared together are studied, however, the degree of concordance for schizophrenia is very much lower—only about 15 percent.

Schizophrenia thus clearly develops much more easily in some genotypes than in others; this indicates a strong hereditary predisposition to the development of the trait. Schizophrenia also serves as a good example of the influence of environmental factors, since concordance for the condition does not appear in 100 percent of identical twins.

Studies of concordance and discordance between identical and fraternal twins have been carried out for many other human characteristics. It has, for example, been known for many years that tuberculosis is a bacterial infection of environmental origin. Yet identical twins raised in the same home show concordance for the disease far more often than do fraternal twins. This finding seems to be explained by the high degree of genetic similarity between the identical twins. While the tuberculosis germ is not inherited, heredity does seem to make one more (or less) susceptible to this particular infection. Thus, the genes of one individual may provide the chemical basis for susceptibility to a disease, while the genes of another may fail to do so.

Indeed, there seem to be genetic differences between disease germs themselves that result in differences in their virulence. Thus, whether a genetically susceptible person actually develops a disease also depends in part on the heredity of the particular strain of bacteria or virus with which he or she must cope. Consequently, unless environmental factors such as these are adequately evaluated, the conclusions drawn from susceptibility studies can be unfortunately misleading.

The above discussion should help to make clear the limits of genetic determinism. The expression of the genotype can always be modified by the environment. It can be argued that all human illnesses have a genetic component and that the basis of all medical therapy is environmental modification. Specifically, this is the hope for the management of genetic diseases. The more that can be learned about the basic molecular and cellular dysfunctions associated with such diseases, the more amenable they will be to environmental manipulation.

Molecular Genetics

Molecular genetics is the study of the processes whereby biological information is stored, copied, repaired and decoded to create protein and other molecules within cells and tissues. The primary focus of molecular genetics is the molecular-level study of the function and structure of genes. Molecular geneticists study the evolution and inheritance of genes. They are curious about how genes control the development and function of organisms. Molecular genetics has lead to mapping the human genome to identify disease-related genetic abnormalities, analyzing genetic changes in tumors and developing new procedures and tests. Genes are the focus, but molecular genetics also involves developmental, cellular and molecular biology.

The term molecular genetics is now redundant because contemporary genetics is thoroughly molecular. Genetics is not made up of two sciences, one molecular and one non-molecular. Nevertheless, practicing biologists still use the term. When they do, they are typically referring to a set of laboratory techniques aimed at identifying and/ or manipulating DNA segments involved in the synthesis of important biological molecules. Scientists often talk and write about the application of these techniques across a broad swath of biomedical sciences. For them, molecular genetics is an investigative approach that involves the application of laboratory methods and research strategies. This approach presupposes basic knowledge about the expression and regulation of genes at the molecular level.

Technique

Amplification

Gene amplification is a procedure in which a certain gene or DNA sequence is replicated many times in a process called DNA replication.

Polymerase Chain Reaction

The main genetic components of the polymerase chain reaction (PCR) are DNA nucleotides, template DNA, primers and Taq polymerase. DNA nucleotides make up the DNA template strand for the specific sequence being amplified and primers are short strands of complementary nucleotides where DNA replication starts. Taq polymerase is a heat stable enzyme that jump-starts the production of new DNA at the high temperatures needed for reaction.

Cloning DNA in Bacteria

Cloning is the process of creating many identical copies of a sequence of DNA. The target DNA sequence is inserted into a cloning vector. Because this vector originates from a self-replicating virus, plasmid, or higher organism cell, when the appropriate size DNA is inserted, the "target and vector DNA fragments are then ligated" to create a recombinant DNA molecule.

The recombinant DNA molecule is then inserted into a bacterial strain (usually *E. coli*) which produces several identical copies of the selected sequence it absorbed through transformation (the mechanism by which bacteria uptake foreign DNA from the environment into their genomes).

Separation and Detection

In separation and detection, DNA and mRNA are isolated from cells and then detected simply by the isolation. Cell cultures are also grown to provide a constant supply of cells ready for isolation.

Cell Cultures

A cell culture for molecular genetics is a culture that is grown in artificial conditions. Some cell types grow well in cultures such as skin cells, but other cells are not as productive in cultures. There are different techniques for each type of cell, some only recently being found to foster growth in stem and nerve cells. Cultures for molecular genetics are frozen in order to preserve all copies of the gene specimen and thawed only when needed. This allows for a steady supply of cells.

DNA Isolation

DNA isolation extracts DNA from a cell in a pure form. First, the DNA is separated from cellular components such as proteins, RNA, and lipids. This is done by placing the chosen cells in a tube with a solution that mechanically, chemically, breaks the cells open. This solution contains enzymes, chemicals, and salts that breaks down the cells except for the DNA. It contains enzymes to dissolve proteins, chemicals to destroy all RNA present, and salts to help pull DNA out of the solution. Next, the DNA is separated from the solution by being spun in a centrifuge, which allows the DNA to collect in the bottom of the tube. After this cycle in the centrifuge the solution is poured off and the DNA is resuspended in a second solution that makes the DNA easy to work with in the future. This results in a concentrated DNA sample that contains thousands of copies of each gene. For large scale projects such as sequencing the human genome, all this work is done by robots.

mRNA Isolation

Expressed DNA that codes for the synthesis of a protein is the final goal for scientists and this expressed DNA is obtained by isolating mRNA (Messenger RNA).

First, laboratories use a normal cellular modification of mRNA that adds up to 200 adenine nucleotides to the end of the molecule (poly(A) tail). Once this has been added, the cell is ruptured and its cell contents are exposed to synthetic beads that are coated with thymine string nucleotides. Because Adenine and Thymine pair together in DNA, the poly(A) tail and synthetic beads are attracted to one another, and once they bind in this process the cell components can be washed away without removing the mRNA. Once the mRNA has been isolated, reverse transcriptase is employed to convert it to single-stranded DNA, from which a stable double-stranded DNA is produced using DNA polymerase. Complementary DNA (cDNA) is much more stable than mRNA and so, once the double-stranded DNA has been produced it represents the expressed DNA sequence scientists look for.

Genetic Screens

Forward Genetics

This technique is used to identify which genes or genetic mutations produce a certain phenotype. A mutagen is very often used to accelerate this process. Once mutants have been isolated, the mutated genes can be molecularly identified.

Forward saturation genetics is a method for treating organisms with a mutagen, then screens the organism's offspring for particular phenotypes. This type of genetic screening is used to find and identify all the genes involved in a trait.

Reverse Genetics

Reverse genetics determines the phenotype that results from a specifically engineered gene. In some organisms, such as yeast and mice, it is possible to induce the deletion of a particular gene, creating what's known as a gene "knockout" - the laboratory origin of so-called "knockout mice" for further study. In other words this process involves the creation of transgenic organisms that do not express a gene of interest. Alternative methods of reverse genetic research include the random induction of DNA deletions and subsequent selection for deletions in a gene of interest, as well as the application of RNA interference.

Gene Therapy

A mutation in a gene can cause encoded proteins and the cells that rely on those proteins to malfunction. Conditions related to gene mutations are called genetic disorders. However, altering a patient's genes can sometimes be used to treat or cure a disease as well. Gene therapy can be used to replace a mutated gene with the correct copy of the gene, to inactivate or knockout the expression of a malfunctioning gene, or to introduce a foreign gene to the body to help fight disease. Major diseases that can be treated with gene therapy include viral infections, cancers, and inherited disorders, including immune system disorders.

Gene therapy delivers a copy of the missing, mutated, or desired gene via a modified virus or vector to the patient's target cells so that a functional form of the protein can then be produced and incorporated into the body. These vectors are often siRNA. Treatment can be either in vivo or ex vivo. The therapy has to be repeated several times for the infected patient to continually be relieved, as repeated cell division and cell death slowly reduces the body's ratio of functional-to-mutant genes. Gene therapy is an appealing alternative to some drug-based approaches, because gene therapy repairs the underlying genetic defect using the patients own cells with minimal side effects. Gene therapies are still in development and mostly used in research settings. All experiments and products are controlled by the U.S. FDA and the NIH.

Classical gene therapies usually require efficient transfer of cloned genes into the disease cells so that the introduced genes are expressed at sufficiently high levels to change the patient's physiology. There are several different physicochemical and biological methods that can be used to transfer genes into human cells. The size of the DNA fragments that can be transferred is very limited, and often the transferred gene is not a conventional gene. Horizontal gene transfer is the transfer of genetic material from one cell to another that is not its offspring. Artificial horizontal gene transfer is a form of genetic engineering.

Applications of Molecular Genetics

Real-World Applications

Population structure – We can quantify genetic differences between populations and, consequently, the level of gene flow and movement among populations throughout a species' range.

Population health and history – Current genetic diversity and historical population bottlenecks can be assessed through the analysis of genetic and genomic data.

Assess taxonomic uncertainty – Information on genetic distinctiveness, along with other lines of evidence such as morphological and behavioral characteristics, can be used to identify and potentially redefine the existing taxonomic classification of a given species or subspecies. Such delineations are highly relevant for species status determinations (endangered, threatened, or at-risk).

Scientists are researching the effects of environmental change on crucial wildlife habitat in the South Central U.S.

Assess adaptive potential – With advancing genomic technology, it is now possible to locate and assess genes that may contribute to a species' ability to respond to environmental change (for example, climate change).

Landscape genetics – Genetic data can be used in conjunction with environmental data (for example, habitat, elevation, roads) to identify landscape features that function as barriers to movement for species.

Assess family relations/mating systems – Mating systems and parentage can be inferred through "genetic fingerprinting" of family members.

Wild horses running on the prairie.

Population modeling – Models can make use of genetic data to predict potential impacts of different management scenarios (for example, fertility control) on the genetic diversity of managed populations.

Estimate population size and survival rates – DNA from non-invasively sampled individuals (using feathers, feces, or hair, for example) can be used as a molecular tag and analyzed with traditional mark-recapture techniques to estimate population size and survival rates.

Species identification – Molecular genetic tests can be used to identify the species and sometimes even the population of origin from a sample (feather, tissue, feces, hair) of unknown origin.

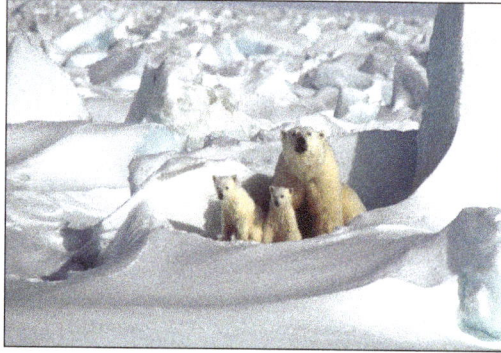

Polar bear mother and two cubs on the Beaufort Sea ice.

Tracking migratory animals – Genetic differences between breeding populations may be used as a basis for assigning migratory animals (like birds, fish, and sea turtles) to a particular breeding population, even if animals are captured on migration or at another location far away from their breeding area.

Gender identification – The gender of an individual can be determined by isolating and analyzing molecular markers on the sex chromosomes; sometimes this is necessary when morphological or behavioral characteristics between males and females are indistinguishable.

A male sage-grouse struts his stuff on his native sage steppe.

Species detection – The presence of difficult-to-observe species can be inferred in an area by screening DNA in environmental samples (eDNA), like water and soil, for species-specific genetic markers.

Black bear cubs.

Dietary analysis – Feces or stomach contents can be screened to figure out what a particular animal has been eating.

References

- Genetics, science: britannica.com, Retrieved 2 April, 2019

- Human-genetics, science: britannica.com, Retrieved 12 June, 2019

- Molecular-genetics: stanford.edu, Retrieved 23 January, 2019

- Science-center-objects, real-world-applications-molecular-genetics, science: usgs.gov, Retrieved 3 February, 2019

2

Nucleic Acid Structure and Gene Expression

The structure of nucleic acids such as DNA and RNA is known as nucleic acid structure. Gene expression is the process that uses the information from a gene for synthesizing a functional gene product. This chapter closely examines the key concepts of nucleic acids and gene expression such as noncoding DNA, DNA replication and translation to provide an extensive understanding of the subject.

DNA

DNA is a long molecule that contains each person's unique genetic code. It holds the instructions for building the proteins that are essential for our bodies to function.

DNA instructions are passed from parent to child, with roughly half of a child's DNA originating from the father and half from the mother.

Structure

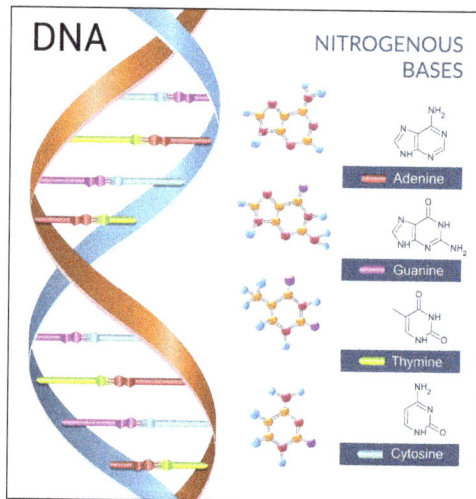

DNA's double helix.

DNA is a two-stranded molecule that appears twisted, giving it a unique shape referred to as the double helix.

Each of the two strands is a long sequence of nucleotides or individual units made of:

- A phosphate molecule,

- A sugar molecule called deoxyribose, containing five carbons,

- A nitrogen-containing region.

There are four types of nitrogen-containing regions called bases:

- Adenine (A),

- Cytosine (C),

- Guanine (G),

- Thymine (T).

The order of these four bases forms the genetic code, which is our instructions for life.

The bases of the two strands of DNA are stuck together to create a ladder-like shape. Within the ladder, A always sticks to T, and G always sticks to C to create the "rungs." The length of the ladder is formed by the sugar and phosphate groups.

Packaging DNA: Chromatin and Chromosomes

Most DNA lives in the nuclei of cells and some is found in mitochondria, which are the powerhouses of the cells.

Because we have so much DNA (2 meters in each cell) and our nuclei are so small, DNA has to be packaged incredibly neatly.

Strands of DNA are looped, coiled and wrapped around proteins called histones. In this coiled state, it is called chromatin.

Chromatin is further condensed, through a process called supercoiling, and it is then packaged into structures called chromosomes. These chromosomes form the familiar "X" shape.

Each chromosome contains one DNA molecule. Humans have 23 pairs of chromosomes or 46 chromosomes in total. Interestingly, fruit flies have 8 chromosomes, and pigeons have 80.

Chromosome 1 is the largest and contains around 8,000 genes. The smallest is chromosome 21 with around 3,000 genes.

DNA Creation of Proteins

For genes to create a protein, there are two main steps:

- Transcription: The DNA code is copied to create messenger RNA (mRNA). RNA is a copy of DNA, but it is normally single-stranded. Another difference is that RNA does not contain the base thymine (T), which is replaced by uracil (U).

- Translation: The mRNA is translated into amino acids by transfer RNA (tRNA).

mRNA is read in three-letter sections called codons. Each codon codes for a specific amino acid or building block of a protein. For instance, the codon GUG codes for the amino acid valine.

There are 20 possible amino acids.

Noncoding DNA

Only about 1 percent of DNA is made up of protein-coding genes; the other 99 percent is noncoding. Noncoding DNA does not provide instructions for making proteins. Scientists once thought noncoding DNA was "junk," with no known purpose. However, it is becoming clear that at least some of it is integral to the function of cells, particularly the control of gene activity. For example, noncoding DNA contains sequences that act as regulatory elements, determining when and where genes are turned on and off. Such elements provide sites for specialized proteins (called transcription factors) to attach (bind) and either activate or repress the process by which the information from genes is turned into proteins (transcription). Noncoding DNA contains many types of regulatory elements:

- Promoters provide binding sites for the protein machinery that carries out transcription. Promoters are typically found just ahead of the gene on the DNA strand.

- Enhancers provide binding sites for proteins that help activate transcription. Enhancers can be found on the DNA strand before or after the gene they control, sometimes far away.

- Silencers provide binding sites for proteins that repress transcription. Like enhancers, silencers can be found before or after the gene they control and can be some distance away on the DNA strand.

- Insulators provide binding sites for proteins that control transcription in a number of ways. Some prevent enhancers from aiding in transcription (enhancer-blocker insulators). Others prevent structural changes in the DNA that repress gene activity (barrier insulators). Some insulators can function as both an enhancer blocker and a barrier.

Other regions of noncoding DNA provide instructions for the formation of certain kinds of RNA molecules. RNA is a chemical cousin of DNA. Examples of specialized RNA molecules produced from noncoding DNA include transfer RNAs (tRNAs) and ribosomal RNAs (rRNAs), which help assemble protein building blocks (amino acids) into a chain that forms a protein; microRNAs (miRNAs), which are short lengths of RNA that block the process of protein production; and long noncoding RNAs (lncRNAs), which are longer lengths of RNA that have diverse roles in regulating gene activity.

Some structural elements of chromosomes are also part of noncoding DNA. For example, repeated noncoding DNA sequences at the ends of chromosomes form telomeres. Telomeres protect the ends of chromosomes from being degraded during the copying of genetic material. Repetitive noncoding DNA sequences also form satellite DNA, which is a part of other structural elements. Satellite DNA is the basis of the centromere, which is the constriction point of the X-shaped chromosome pair. Satellite DNA also forms heterochromatin, which is densely packed DNA that is important for controlling gene activity and maintaining the structure of chromosomes.

Some noncoding DNA regions, called introns, are located within protein-coding genes but are removed before a protein is made. Regulatory elements, such as enhancers, can be located in introns. Other noncoding regions are found between genes and are known as intergenic regions.

DNA Replication

DNA replication uses a semi-conservative method that results in a double-stranded DNA with one parental strand and a new daughter strand.

Watson and Crick's discovery that DNA was a two-stranded double helix provided a hint as to how DNA is replicated. During cell division, each DNA molecule has to be perfectly copied to ensure identical DNA molecules to move to each of the two daughter cells. The double-stranded structure of DNA suggested that the two strands might separate during replication with each strand serving as a template from which the new complementary strand for each is copied, generating two double-stranded molecules from one.

Models of Replication

There were three models of replication possible from such a scheme: conservative, semi-conservative, and dispersive. In conservative replication, the two original DNA strands, known as the parental strands, would re-basepair with each other after being used as templates to synthesize new strands; and the two newly-synthesized strands, known as the daughter strands, would also basepair with each other; one of the two DNA molecules after replication would be "all-old" and the other would be "all-new". In semi-conservative replication, each of the two parental DNA strands would act as a template for new DNA strands to be synthesized, but after replication, each parental

DNA strand would basepair with the complementary newly-synthesized strand just synthesized, and both double-stranded DNAs would include one parental or "old" strand and one daughter or "new" strand. In dispersive replication, after replication both copies of the new DNAs would somehow have alternating segments of parental DNA and newly-synthesized DNA on each of their two strands.

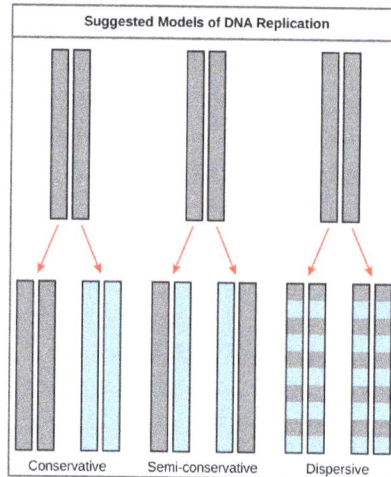

Suggested Models of DNA Replication: The three suggested models of DNA replication. Grey indicates the original parental DNA strands or segments and blue indicates newly-synthesized daughter DNA strands or segments.

To determine which model of replication was accurate, a seminal experiment was performed in 1958 by two researchers: Matthew Meselson and Franklin Stahl.

Meselson and Stahl

Meselson and Stahl were interested in understanding how DNA replicates. They grew E. coli for several generations in a medium containing a "heavy" isotope of nitrogen (^{15}N) that is incorporated into nitrogenous bases and, eventually, into the DNA. The E. coli culture was then shifted into medium containing the common "light" isotope of nitrogen (^{14}N) and allowed to grow for one generation. The cells were harvested and the DNA was isolated. The DNA was centrifuged at high speeds in an ultracentrifuge in a tube in which a cesium chloride density gradient had been established. Some cells were allowed to grow for one more life cycle in ^{14}N and spun again.

During the density gradient ultracentrifugation, the DNA was loaded into a gradient (Meselson and Stahl used a gradient of cesium chloride salt, although other materials such as sucrose can also be used to create a gradient) and spun at high speeds of 50,000 to 60,000 rpm. In the ultracentrifuge tube, the cesium chloride salt created a density gradient, with the cesium chloride solution being more dense the farther down the tube you went. Under these circumstances, during the spin the DNA was pulled down the ultracentrifuge tube by centrifugal force until it arrived at the spot in the salt gradient where the DNA molecules' density matched that of the surrounding salt solution. At the point, the molecules stopped sedimenting and formed a stable band. By looking at the relative positions of bands of molecules run in the same gradients, you can determine the relative densities of different molecules. The molecules that form the lowest bands have the highest densities.

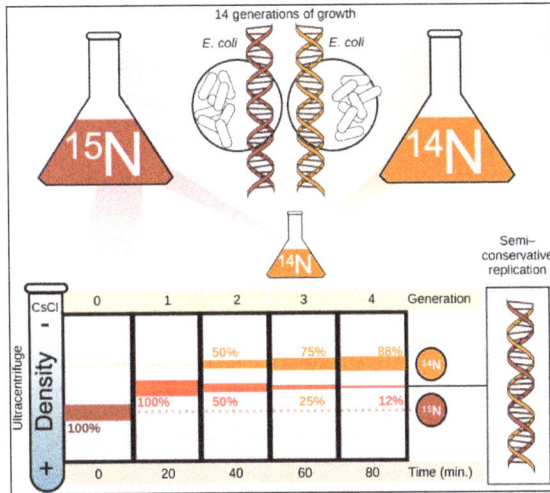

Meselson and Stahl

Meselson and Stahl experimented with E. coli grown first in heavy nitrogen (^{15}N) then in ligher nitrogen (^{14}N.) DNA grown in ^{15}N (red band) is heavier than DNA grown in ^{14}N (orange band) and sediments to a lower level in the cesium chloride density gradient in an ultracentrifuge. When DNA grown in ^{15}N is switched to media containing ^{14}N, after one round of cell division the DNA sediments halfway between the ^{15}N and ^{14}N levels, indicating that it now contains fifty percent ^{14}N and fifty percent ^{15}N. In subsequent cell divisions, an increasing amount of DNA contains ^{14}N only. These data support the semi-conservative replication model.

DNA from cells grown exclusively in ^{15}N produced a lower band than DNA from cells grown exclusively in ^{14}N. So DNA grown in ^{15}N had a higher density, as would be expected of a molecule with a heavier isotope of nitrogen incorporated into its nitrogenous bases. Meselson and Stahl noted that after one generation of growth in ^{14}N (after cells had been shifted from ^{15}N), the DNA molecules produced only single band intermediate in position in between DNA of cells grown exclusively in ^{15}N and DNA of cells grown exclusively in ^{14}N. This suggested either a semi-conservative or dispersive mode of replication. Conservative replication would have resulted in two bands; one representing the parental DNA still with exclusively ^{15}N in its nitrogenous bases and the other representing the daughter DNA with exclusively ^{14}N in its nitrogenous bases. The single band actually seen indicated that all the DNA molecules contained equal amounts of both ^{15}N and ^{14}N.

The DNA harvested from cells grown for two generations in ^{14}N formed two bands: one DNA band was at the intermediate position between ^{15}N and ^{14}N and the other corresponded to the band of exclusively ^{14}N DNA. These results could only be explained if DNA replicates in a semi-conservative manner. Dispersive replication would have resulted in exclusively a single band in each new generation, with the band slowly moving up closer to the height of the ^{14}N DNA band. Therefore, dispersive replication could also be ruled out.

Meselson and Stahl's results established that during DNA replication, each of the two strands that make up the double helix serves as a template from which new strands are synthesized. The new strand will be complementary to the parental or "old" strand and the new strand will remain basepaired to the old strand. So each "daughter" DNA actually consists of one "old" DNA strand and one newly-synthesized strand. When two daughter DNA copies are formed, they have the identical sequences to one another and identical sequences to the original parental DNA, and the two daughter DNAs are divided equally into the two daughter cells, producing daughter cells that are genetically identical to one another and genetically identical to the parent cell.

DNA Replication in Prokaryotes

Prokaryotic DNA is replicated by DNA polymerase III in the 5′ to 3′ direction at a rate of 1000 nucleotides per second.

DNA replication employs a large number of proteins and enzymes, each of which plays a critical role during the process. One of the key players is the enzyme DNA polymerase, which adds nucleotides one by one to the growing DNA chain that are complementary to the template strand. The addition of nucleotides requires energy; this energy is obtained from the nucleotides that have three phosphates attached to them, similar to ATP which has three phosphate groups attached. When the bond between the phosphates is broken, the energy released is used to form the phosphodiester bond between the incoming nucleotide and the growing chain. In prokaryotes, three main types of polymerases are known: DNA pol I, DNA pol II, and DNA pol III. DNA pol III is the enzyme required for DNA synthesis; DNA pol I and DNA pol II are primarily required for repair.

There are specific nucleotide sequences called origins of replication where replication begins. In *E. coli*, which has a single origin of replication on its one chromosome (as do most prokaryotes), it is approximately 245 base pairs long and is rich in AT sequences. The origin of replication is recognized by certain proteins that bind to this site. An enzyme called helicase unwinds the DNA by breaking the hydrogen bonds between the nitrogenous base pairs. ATP hydrolysis is required for this process. As the DNA opens up, Y-shaped structures called replication forks are formed. Two replication forks at the origin of replication are extended bi-directionally as replication proceeds. Single-strand binding proteins coat the strands of DNA near the replication fork to prevent the single-stranded DNA from winding back into a double helix. DNA polymerase is able to add nucleotides only in the 5′ to 3′ direction (a new DNA strand can be extended only in this direction). It also requires a free 3'-OH group to which it can add nucleotides by forming a phosphodiester bond between the 3'-OH end and the 5' phosphate of the next nucleotide. This means that it cannot add nucleotides if a free 3'-OH group is not available. Another enzyme, RNA primase, synthesizes an RNA primer that is about five to ten nucleotides long and complementary to the DNA, priming DNA synthesis. A primer provides the free 3'-OH end to start replication. DNA polymerase then extends this RNA primer, adding nucleotides one by one that are complementary to the template strand.

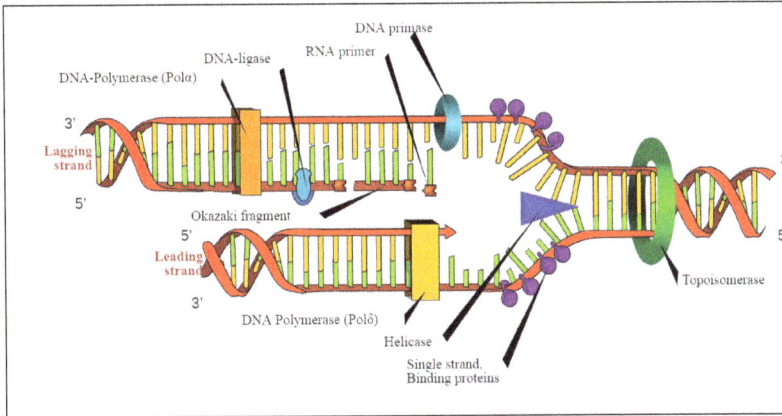

DNA Replication in Prokaryotes.

A replication fork is formed when helicase separates the DNA strands at the origin of replication. The DNA tends to become more highly coiled ahead of the replication fork. Topoisomerase breaks and reforms DNA's phosphate backbone ahead of the replication fork, thereby relieving the pressure that results from this supercoiling. Single-strand binding proteins bind to the single-stranded DNA to prevent the helix from re-forming. Primase synthesizes an RNA primer. DNA polymerase III uses this primer to synthesize the daughter DNA strand. On the leading strand, DNA is synthesized continuously, whereas on the lagging strand, DNA is synthesized in short stretches called Okazaki fragments. DNA polymerase I replaces the RNA primer with DNA. DNA ligase seals the gaps between the Okazaki fragments, joining the fragments into a single DNA molecule.

The replication fork moves at the rate of 1000 nucleotides per second. DNA polymerase can only extend in the 5' to 3' direction, which poses a slight problem at the replication fork. As we know, the DNA double helix is anti-parallel; that is, one strand is in the 5' to 3' direction and the other is oriented in the 3' to 5' direction. One strand (the leading strand), complementary to the 3' to 5' parental DNA strand, is synthesized continuously towards the replication fork because the polymerase can add nucleotides in this direction. The other strand (the lagging strand), complementary to the 5' to 3' parental DNA, is extended away from the replication fork in small fragments known as Okazaki fragments, each requiring a primer to start the synthesis. Okazaki fragments are named after the Japanese scientist who first discovered them.

The leading strand can be extended by one primer alone, whereas the lagging strand needs a new primer for each of the short Okazaki fragments. The overall direction of the lagging strand will be 3′ to 5′, while that of the leading strand will be 5′ to 3′. The sliding clamp (a ring-shaped protein that binds to the DNA) holds the DNA polymerase in place as it continues to add nucleotides. Topoisomerase prevents the over-winding of the DNA double helix ahead of the replication fork as the DNA is opening up; it does so by causing temporary nicks in the DNA helix and then resealing it. As synthesis proceeds, the RNA primers are replaced by DNA. The primers are removed by the exonuclease activity of DNA pol I, while the gaps are filled in by deoxyribonucleotides. The nicks that remain between the newly-synthesized DNA (that replaced the RNA primer) and the previously-synthesized DNA are sealed by the enzyme DNA ligase that catalyzes the formation of phosphodiester linkage between the 3′-OH end of one nucleotide and the 5′ phosphate end of the other fragment.

The table summarizes the enzymes involved in prokaryotic DNA replication and the functions of each.

Prokaryotic DNA Replication : Enzymes and their Function	
Enzyme/protein	Specific Function
DNA pol I	Exonuclease activity remove RNA primer and replaces with newly synthesized DNA.
DNA pol II	Repair function.
DNA pol III	Main enzyme that adds nucleotides in the 5'-3' direction.
Helicase	Opens the DNA helix by breaking hydrogen bonds between the nitrogenous bases.
Ligase	Seals the gaps between the Okazaki fragments to create one continuous DNA strand.
Primase	Synthesizes RNA primers needed to start replication.
Sliding Clamp	Helps to hold the DNA polymerase in place when nucleotides are being added.
Topoisomerase	Helps relieve the stress on DNA when unwinding by causing breaks and then resealing the DNA.
Single-strand binding proteins(SSB)	Binds to single-stranded DNA to avoid DNA rewinding back.

Prokaryotic DNA Replication: Enzymes and their Function: The enzymes involved in prokaryotic DNA replication and their functions are summarized on this table.

DNA Replication in Eukaryotes

DNA replication in eukaryotes occurs in three stages: initiation, elongation, and termination, which are aided by several enzymes.

Because eukaryotic genomes are quite complex, DNA replication is a very complicated process that involves several enzymes and other proteins. It occurs in three main stages: initiation, elongation, and termination.

Initiation

Eukaryotic DNA is bound to proteins known as histones to form structures called nucleosomes. During initiation, the DNA is made accessible to the proteins and enzymes involved in the replication process. There are specific chromosomal locations called origins of replication where replication begins. In some eukaryotes, like yeast, these locations are defined by having a specific sequence of basepairs to which the replication initiation proteins bind. In other eukaryotes, like humans, there does not appear to be a consensus sequence for their origins of replication. Instead, the replication initiation proteins might identify and bind to specific modifications to the nucleosomes in the origin region.

Certain proteins recognize and bind to the origin of replication and then allow the other proteins necessary for DNA replication to bind the same region. The first proteins to bind the DNA are said to "recruit" the other proteins. Two copies of an enzyme called helicase are among the proteins recruited to the origin. Each helicase unwinds and separates the DNA helix into single-stranded DNA. As the DNA opens up, Y-shaped structures called replication forks are formed. Because two helicases bind, two replication forks are formed at the origin of replication; these are extended in both directions as replication proceeds creating a replication bubble. There are multiple origins of replication on the eukaryotic chromosome which allow replication to occur simultaneously in hundreds to thousands of locations along each chromosome.

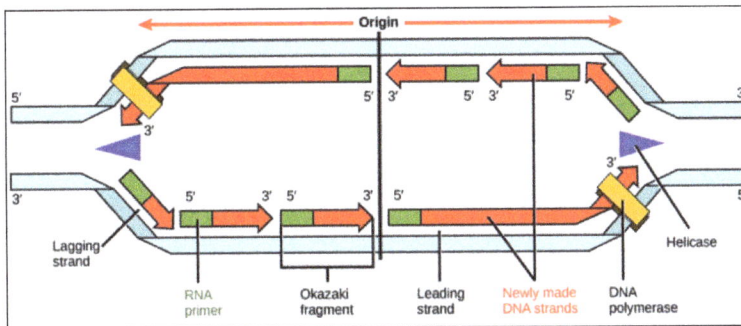

Replication Fork Formation.

A replication fork is formed by the opening of the origin of replication; helicase separates the DNA strands. An RNA primer is synthesized by primase and is elongated by the DNA polymerase. On the leading strand, only a single RNA primer is needed, and DNA is synthesized continuously, whereas on the lagging strand, DNA is synthesized in short stretches, each of which must start with its own RNA primer. The DNA fragments are joined by DNA ligase.

Elongation

During elongation, an enzyme called DNA polymerase adds DNA nucleotides to the 3′ end of the newly synthesized polynucleotide strand. The template strand specifies which of the four DNA nucleotides (A, T, C, or G) is added at each position along the new chain. Only the nucleotide complementary to the template nucleotide at that position is added to the new strand.

DNA polymerase contains a groove that allows it to bind to a single-stranded template DNA and travel one nucleotide at at time. For example, when DNA polymerase meets an adenosine nucleotide on the template strand, it adds a thymidine to the 3' end of the newly synthesized strand, and then moves to the next nucleotide on the template strand. This process will continue until the DNA polymerase reaches the end of the template strand.

DNA polymerase cannot initiate new strand synthesis; it only adds new nucleotides at the 3' end of an existing strand. All newly synthesized polynucleotide strands must be initiated by a specialized RNA polymerase called primase. Primase initiates polynucleotide synthesis and by creating a short RNA polynucleotide strand complementary to template DNA strand. This short stretch of RNA nucleotides is called the primer. Once RNA primer has been synthesized at the template DNA, primase exits, and DNA polymerase extends the new strand with nucleotides complementary to the template DNA.

Eventually, the RNA nucleotides in the primer are removed and replaced with DNA nucleotides. Once DNA replication is finished, the daughter molecules are made entirely of continuous DNA nucleotides, with no RNA portions.

The Leading and Lagging Strands

DNA polymerase can only synthesize new strands in the 5' to 3' direction. Therefore, the two newly-synthesized strands grow in opposite directions because the template strands at each replication fork are antiparallel. The "leading strand" is synthesized continuously toward the replication fork as helicase unwinds the template double-stranded DNA.

The "lagging strand" is synthesized in the direction away from the replication fork and away from the DNA helicase unwinds. This lagging strand is synthesized in pieces because the DNA polymerase can only synthesize in the 5' to 3' direction, and so it constantly encounters the previously-synthesized new strand. The pieces are called Okazaki fragments, and each fragment begins with its own RNA primer.

Termination

Eukaryotic chromosomes have multiple origins of replication, which initiate replication almost simultaneously. Each origin of replication forms a bubble of duplicated DNA on either side of the origin of replication. Eventually, the leading strand of one replication bubble reaches the lagging strand of another bubble, and the lagging strand will reach the 5' end of the previous Okazaki fragment in the same bubble.

DNA polymerase halts when it reaches a section of DNA template that has already been replicated. However, DNA polymerase cannot catalyze the formation of a phosphodiester bond between the two segments of the new DNA strand, and it drops off. These unattached sections of the sugar-phosphate backbone in an otherwise full-replicated DNA strand are called nicks.

Once all the template nucleotides have been replicated, the replication process is not yet over. RNA primers need to be replaced with DNA, and nicks in the sugar-phosphate backbone need to be connected.

The group of cellular enzymes that remove RNA primers include the proteins FEN1 (flap endonulcease 1) and RNase H. The enzymes FEN1 and RNase H remove RNA primers at the start of each leading strand and at the start of each Okazaki fragment, leaving gaps of unreplicated template DNA. Once the primers are removed, a free-floating DNA polymerase lands at the 3' end of the preceding DNA fragment and extends the DNA over the gap. However, this creates new nicks (unconnected sugar-phosphate backbone).

In the final stage of DNA replication, the enyzme ligase joins the sugar-phosphate backbones at each nick site. After ligase has connected all nicks, the new strand is one long continuous DNA strand, and the daughter DNA molecule is complete.

Telomere Replication

As DNA polymerase alone cannot replicate the ends of chromosomes, telomerase aids in their replication and prevents chromosome degradation.

The End Problem of Linear DNA Replication

Linear chromosomes have an end problem. After DNA replication, each newly synthesized DNA strand is shorter at its 5' end than at the parental DNA strand's 5' end. This produces a 3' overhang at one end (and one end only) of each daughter DNA strand, such that the two daughter DNAs have their 3' overhangs at opposite ends.

Every RNA primer synthesized during replication can be removed and replaced with DNA strands except the RNA primer at the 5' end of the newly synthesized strand. This small section of RNA can only be removed, not replaced with DNA. Enzymes RNase H and FEN1 remove RNA primers, but DNA Polymerase will add new DNA only if the DNA Polymerase has an existing strand 5' to it ("behind" it) to extend. However, there is no more DNA in the 5' direction after the final RNA primer, so DNA polymerse cannot replace the RNA with DNA. Therefore, both daughter DNA strands have an incomplete 5' strand with 3' overhang.

In the absence of additional cellular processes, nucleases would digest these single-stranded 3' overhangs. Each daughter DNA would become shorter than the parental

DNA, and eventually entire DNA would be lost. To prevent this shortening, the ends of linear eukaryotic chromosomes have special structures called telomeres.

Telomere Replication

The ends of the linear chromosomes are known as telomeres: repetitive sequences that code for no particular gene. These telomeres protect the important genes from being deleted as cells divide and as DNA strands shorten during replication.

In humans, a six base pair sequence, TTAGGG, is repeated 100 to 1000 times. After each round of DNA replication, some telomeric sequences are lost at the 5′ end of the newly synthesized strand on each daughter DNA, but because these are noncoding sequences, their loss does not adversely affect the cell. However, even these sequences are not unlimited. After sufficient rounds of replication, all the telomeric repeats are lost, and the DNA risks losing coding sequences with subsequent rounds.

The discovery of the enzyme telomerase helped in the understanding of how chromosome ends are maintained. The telomerase enzyme attaches to the end of a chromosome and contains a catalytic part and a built-in RNA template. Telomerase adds complementary RNA bases to the 3′ end of the DNA strand. Once the 3′ end of the lagging strand template is sufficiently elongated, DNA polymerase adds the complementary nucleotides to the ends of the chromosomes; thus, the ends of the chromosomes are replicated.

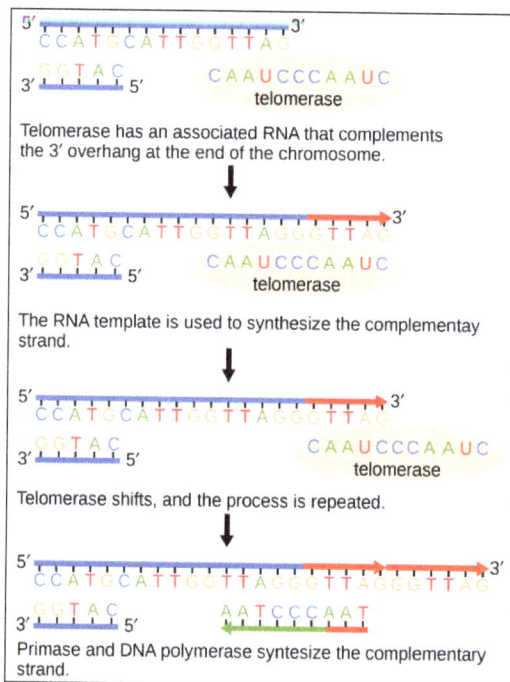

Telomerase is important for maintaining chromosome integrity: The ends of linear chromosomes are maintained by the action of the telomerase enzyme.

Telomerase and Aging

Telomerase is typically active in germ cells and adult stem cells, but is not active in adult somatic cells. As a result, telomerase does not protect the DNA of adult somatic cells and their telomeres continually shorten as they undergo rounds of cell division.

Translation

The genes in DNA encode protein molecules, which are the "workhorses" of the cell, carrying out all the functions necessary for life. For example, enzymes, including those that metabolize nutrients and synthesize new cellular constituents, as well as DNA polymerases and other enzymes that make copies of DNA during cell division, are all proteins.

In the simplest sense, expressing a gene means manufacturing its corresponding protein, and this multilayered process has two major steps. In the first step, the information in DNA is transferred to a messenger RNA (mRNA) molecule by way of a process called transcription. During transcription, the DNA of a gene serves as a template for complementary base-pairing, and an enzyme called RNA polymerase II catalyzes the formation of a pre-mRNA molecule, which is then processed to form mature mRNA. The resulting mRNA is a single-stranded copy of the gene, which next must be translated into a protein molecule.

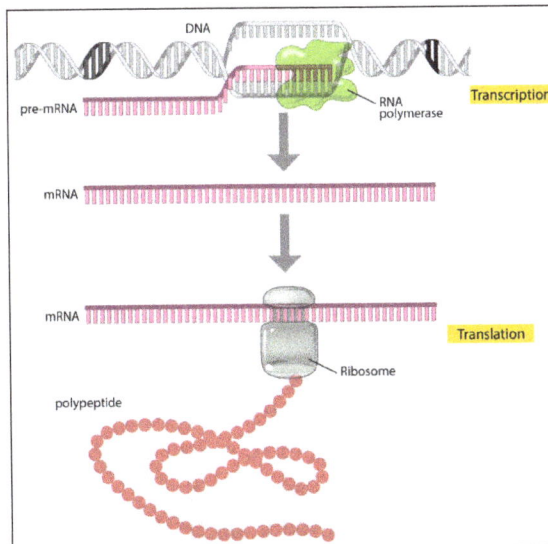

A gene is expressed through the processes of transcription and translation. During transcription, the enzyme RNA polymerase (green) uses DNA as a template to produce a pre-mRNA transcript (pink). The pre-mRNA is processed to form a mature mRNA molecule that can be translated to build the protein molecule (polypeptide) encoded by the original gene.

During translation, which is the second major step in gene expression, the mRNA is "read" according to the genetic code, which relates the DNA sequence to the amino acid sequence in proteins. Each group of three bases in mRNA constitutes a codon, and each codon specifies a particular amino acid (hence, it is a triplet code). The mRNA sequence is thus used as a template to assemble—in order—the chain of amino acids that form a protein.

The amino acids specified by each mRNA codon.
Multiple codons can code for the same amino acid.

Occurance of Translation

Within all cells, the translation machinery resides within a specialized organelle called the ribosome. In eukaryotes, mature mRNA molecules must leave the nucleus and travel to the cytoplasm, where the ribosomes are located. On the other hand, in prokaryotic organisms, ribosomes can attach to mRNA while it is still being transcribed. In this situation, translation begins at the 5' end of the mRNA while the 3' end is still attached to DNA.

In all types of cells, the ribosome is composed of two subunits: the large (50S) subunit and the small (30S) subunit (S, for svedberg unit, is a measure of sedimentation velocity and, therefore, mass). Each subunit exists separately in the cytoplasm, but the two join together on the mRNA molecule. The ribosomal subunits contain proteins and specialized RNA molecules—specifically, ribosomal RNA (rRNA) and transfer RNA (tRNA). The tRNA molecules are adaptor molecules—they have one end that can read the triplet code in the mRNA through complementary base-pairing, and another end that attaches to a specific amino acid. The idea that tRNA was an adaptor molecule was first proposed by Francis Crick, co-discoverer of DNA structure, who did much of the key work in deciphering the genetic code.

Within the ribosome, the mRNA and aminoacyl-tRNA complexes are held together closely, which facilitates base-pairing. The rRNA catalyzes the attachment of each new amino acid to the growing chain.

The Beginning of mRNA is not Translated

Interestingly, not all regions of an mRNA molecule correspond to particular amino acids. In particular, there is an area near the 5' end of the molecule that is known as the untranslated region (UTR) or leader sequence. This portion of mRNA is located between the first nucleotide that is transcribed and the start codon (AUG) of the coding region, and it does not affect the sequence of amino acids in a protein.

Purpose of the UTR: It turns out that the leader sequence is important because it contains a ribosome-binding site. In bacteria, this site is known as the Shine-Dalgarno box (AGGAGG), after scientists John Shine and Lynn Dalgarno, who first characterized it. A similar site in vertebrates was characterized by Marilyn Kozak and is thus known as the Kozak box. In bacterial mRNA, the 5' UTR is normally short; in human mRNA, the median length of the 5' UTR is about 170 nucleotides. If the leader is long, it may contain regulatory sequences, including binding sites for proteins, that can affect the stability of the mRNA or the efficiency of its translation.

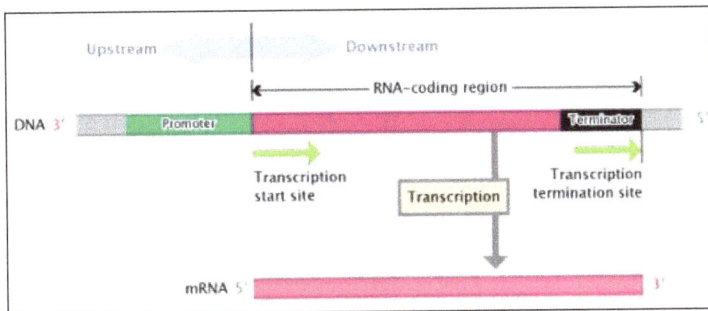

A DNA transcription unit.

A DNA transcription unit is composed, from its 3' to 5' end, of an RNA-coding region (pink rectangle) flanked by a promoter region (green rectangle) and a terminator region (black rectangle). Regions to the left, or moving towards the 3' end, of the transcription start site are considered upstream; regions to the right, or moving towards the 5' end, of the transcription start site are considered downstream.

Translation Begins After the Assembly of a Complex Structure

The translation of mRNA begins with the formation of a complex on the mRNA. First, three initiation factor proteins (known as IF1, IF2, and IF3) bind to the small subunit of the ribosome. This preinitiation complex and a methionine-carrying tRNA then bind to the mRNA, near the AUG start codon, forming the initiation complex.

When translation begins, the small subunit of the ribosome and an initiator tRNA molecule assemble on the mRNA transcript. The small subunit of the ribosome has three binding sites: an amino acid site (A), a polypeptide site (P), and an exit site (E). The initiator tRNA molecule carrying the amino acid methionine binds to the AUG start codon of the mRNA transcript at the ribosome's P site where it will become the first amino acid incorporated into the growing polypeptide chain. Here, the initiator tRNA molecule is shown binding after the small ribosomal subunit has assembled on the mRNA; the order in which this occurs is unique to prokaryotic cells. In eukaryotes, the free initiator tRNA first binds the small ribosomal subunit to form a complex. The complex then binds the mRNA transcript, so that the tRNA and the small ribosomal subunit bind the mRNA simultaneously.

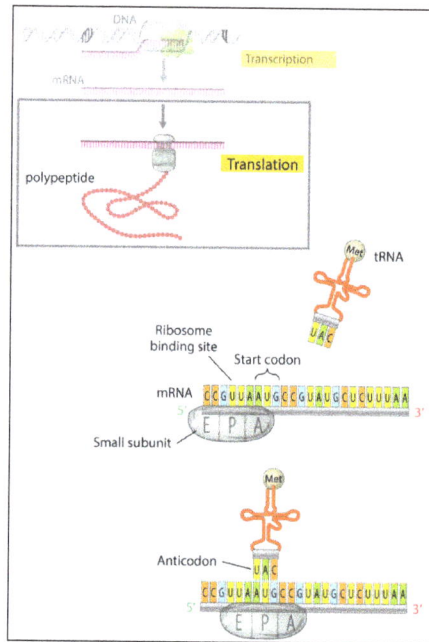

The translation initiation complex.

Although methionine (Met) is the first amino acid incorporated into any new protein, it is not always the first amino acid in mature proteins—in many proteins, methionine is removed after translation. In fact, if a large number of proteins are sequenced and compared with their known gene sequences, methionine (or formylmethionine) occurs at the N-terminus of all of them. However, not all amino acids are equally likely to occur second in the chain, and the second amino acid influences whether the initial methionine is enzymatically removed. For example, many proteins begin with methionine followed by alanine. In both prokaryotes and eukaryotes, these proteins have the methionine removed, so that alanine becomes the N-terminal amino acid. However, if the second amino acid is lysine, which is also frequently the case, methionine is not removed. These proteins therefore begin with methionine followed by lysine.

Table shows the N-terminal sequences of proteins in prokaryotes and eukaryotes, based on a sample of 170 prokaryotic and 120 eukaryotic proteins. In the table, M represents methionine, A represents alanine, K represents lysine, S represents serine, and T represents threonine.

Table: N-Terminal Sequences of Proteins

N-Terminal Sequence	Percent of Prokaryotic Proteins with This Sequence	Percent of Eukaryotic Proteins with This Sequence
MA*	28.24%	19.17%
MK**	10.59%	2.50%
MS*	9.41%	11.67%
MT*	7.65%	6.67%

* Methionine was removed in all of these proteins.

** Methionine was not removed from any of these proteins.

Once the initiation complex is formed on the mRNA, the large ribosomal subunit binds to this complex, which causes the release of IFs (initiation factors). The large subunit of the ribosome has three sites at which tRNA molecules can bind. The A (amino acid) site is the location at which the aminoacyl-tRNA anticodon base pairs up with the mRNA codon, ensuring that correct amino acid is added to the growing polypeptide chain. The P (polypeptide) site is the location at which the amino acid is transferred from its tRNA to the growing polypeptide chain. Finally, the E (exit) site is the location at which the "empty" tRNA sits before being released back into the cytoplasm to bind another amino acid and repeat the process. The initiator methionine tRNA is the only aminoacyl-tRNA that can bind in the P site of the ribosome, and the A site is aligned with the second mRNA codon. The ribosome is thus ready to bind the second aminoacyl-tRNA at the A site, which will be joined to the initiator methionine by the first peptide bond.

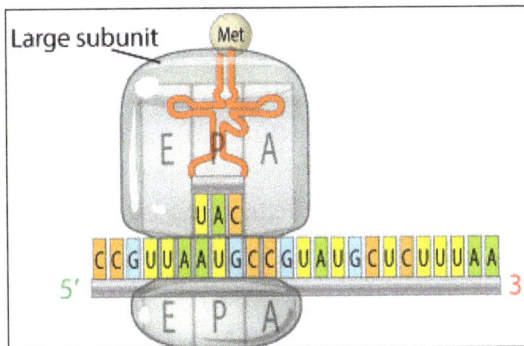

The large ribosomal subunit binds to the small ribosomal subunit to complete the initiation complex.

The initiator tRNA molecule, carrying the methionine amino acid that will serve as the first amino acid of the polypeptide chain, is bound to the P site on the ribosome. The A site is aligned with the next codon, which will be bound by the anticodon of the next incoming tRNA.

The Elongation Phase

The next phase in translation is known as the elongation phase. First, the ribosome moves along the mRNA in the 5'-to-3'direction, which requires the elongation factor G, in a process called translocation. The tRNA that corresponds to the second codon can then bind to the A site, a step that requires elongation factors (in E. coli, these are called EF-Tu and EF-Ts), as well as guanosine triphosphate (GTP) as an energy source for the process. Upon binding of the tRNA-amino acid complex in the A site, GTP is cleaved to form guanosine diphosphate (GDP), then released along with EF-Tu to be recycled by EF-Ts for the next round.

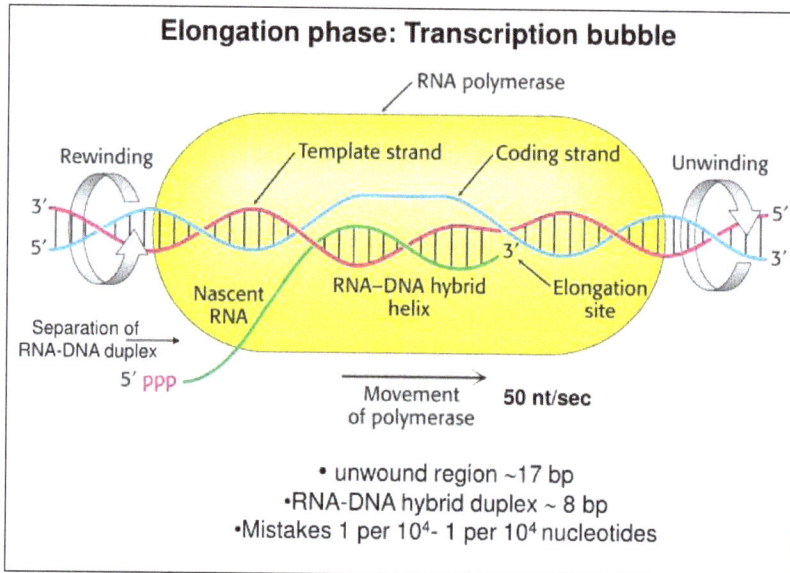

Next, peptide bonds between the now-adjacent first and second amino acids are formed through a peptidyl transferase activity. For many years, it was thought that an enzyme catalyzed this step, but recent evidence indicates that the transferase activity is a catalytic function of rRNA. After the peptide bond is formed, the ribosome shifts, or translocates, again, thus causing the tRNA to occupy the E site. The tRNA is then released to the cytoplasm to pick up another amino acid. In addition, the A site is now empty and ready to receive the tRNA for the next codon.

This process is repeated until all the codons in the mRNA have been read by tRNA molecules, and the amino acids attached to the tRNAs have been linked together in the growing polypeptide chain in the appropriate order. At this point, translation must be terminated, and the nascent protein must be released from the mRNA and ribosome.

Termination of Translation

There are three termination codons that are employed at the end of a protein-coding sequence in mRNA: UAA, UAG, and UGA. No tRNAs recognize these codons. Thus, in the place of these tRNAs, one of several proteins, called release factors, binds and facilitates release of the mRNA from the ribosome and subsequent dissociation of the ribosome.

Comparing Eukaryotic and Prokaryotic Translation

The translation process is very similar in prokaryotes and eukaryotes. Although different elongation, initiation, and termination factors are used, the genetic code is generally identical. As previously noted, in bacteria, transcription and translation take place simultaneously, and mRNAs are relatively short-lived. In eukaryotes, however, mRNAs have highly variable half-lives, are subject to modifications, and must exit the nucleus to be translated; these multiple steps offer additional opportunities to regulate levels of protein production, and thereby fine-tune gene expression.

RNA

RNA stands for Ribonucleic Acid. It is a complex compound of high molecular weight that functions in cellular protein synthesis and replaces DNA (deoxyribonucleic acid) as a carrier of genetic codes in some viruses. RNA consists of ribose nucleotides (nitrogenous bases appended to a ribose sugar) attached by phosphodiester bonds, forming strands of varying lengths. The nitrogenous bases in RNA are adenine, guanine, cytosine, and uracil, which replaces thymine in DNA.

The ribose sugar of RNA is a cyclical structure consisting of five carbons and one oxygen. The presence of a chemically reactive hydroxyl (–OH) group attached to the second carbon group in the ribose sugar molecule makes RNA prone to hydrolysis. This chemical lability of RNA, compared with DNA, which does not have a reactive –OH group in the same position on the sugar moiety (deoxyribose), is thought to be one reason why DNA evolved to be the preferred carrier of genetic information in most organisms. The structure of the RNA molecule was described by R.W. Holley in 1965.

RNA Structure

RNA typically is a single-stranded biopolymer. However, the presence of self-complementary sequences in the RNA strand leads to intrachain base-pairing and folding of the ribonucleotide chain into complex structural forms consisting of bulges and helices. The three-dimensional structure of RNA is critical to its stability and function, allowing the ribose sugar and the nitrogenous bases to be modified in numerous different ways by cellular enzymes that attach chemical groups (e.g., methyl groups) to the chain. Such modifications enable the formation of chemical bonds between distant regions in the RNA strand, leading to complex contortions in the RNA chain, which further stabilizes the RNA structure. Molecules with weak structural modifications and stabilization may be readily destroyed. As an example, in an initiator transfer RNA (tRNA) molecule that lacks a methyl group ($tRNA_i^{Met}$), modification at position 58 of the tRNA chain renders the molecule unstable and hence nonfunctional; the nonfunctional chain is destroyed by cellular tRNA quality control mechanisms.

RNAs can also form complexes with molecules known as ribonucleoproteins (RNPs). The RNA portion of at least one cellular RNP has been shown to act as a biological catalyst, a function previously ascribed only to proteins.

Types of RNA

Of the many types of RNA, the three most well-known and most commonly studied are messenger RNA (mRNA), transfer RNA (tRNA), and ribosomal RNA (rRNA), which are present in all organisms. These and other types of RNAs primarily carry out biochemical reactions, similar to enzymes. Some, however, also have complex regulatory functions in cells. Owing to their involvement in many regulatory processes, to their abundance, and to their diverse functions, RNAs play important roles in both normal cellular processes and diseases.

Ribosomal RNA

Ribosomal ribonucleic acid (rRNA) is the RNA component of ribosomes, the molecular machines that catalyze protein synthesis. Ribosomal RNA constitute over sixty percent of the ribosome by weight and are crucial for all its functions – from binding to mRNA and recruiting tRNA to catalyzing the formation of a peptide bond between two amino acids. Even the structure of a ribosome is determined by the three-dimensional shape of its rRNA core. Proteins present in the ribosome serve to stabilize this structure through interactions with the core.

Ribosomal RNA are transcribed in the nucleus, at specific structures called nucleoli. These are dense, spherical shapes that form around genetic loci coding for rRNA. Nucleoli are also crucial for the eventual biogenesis of ribosomes, through sequestration of ribosomal proteins.

Discovery of Ribosomal RNA

Ribosomal RNA was discovered during cell fractionation experiments investigating the role of RNA viruses in causing cancer. Fractionation is a method where cell membranes are carefully and selectively removed while keeping the function of cellular organelles intact. This homogenized cytoplasm is then centrifuged at increasing speeds so that organelles separate according to density. The initial experiments that revealed the presence of rRNA extracted a fraction that was thought to represent a new sub-cellular organelle, microsome, specializing in protein synthesis. Later, it was seen that it was the presence of ribosomes on endoplasmic reticulum that led to the detection of RNA in these samples.

Since ribosomal subunits and rRNA were first detected through differential centrifugation, they are still identified by their rate of sedimentation, through the Svedberg coefficient. However, since these are not measures of molecular weight, the coefficients cannot be directly added. For example, in prokaryotic ribosomes when the 50S larger subunit and 30S smaller subunit come together, the complex has a Svedberg coefficient of 70S.

Types of Ribosomal RNA

Both prokaryotic and eukaryotic ribosomes are made of a larger and smaller subunit and these two units come together during mRNA translation. The smaller subunit in prokaryotes is made of an RNA molecule about 1500 nucleotides in length with a Svedberg coefficient of 16S. Together with ribosomal proteins, the smaller subunit has a sedimentation rate of 30S. This is paired with the larger subunit, having two RNA molecules – one that is nearly 3000 nucleotides (23S) in length and the other is a short sequence of 120 nucleotides (5S). These RNA molecules are accompanied by proteins that give rise to the larger 50S subunit.

The eukaryotic ribosome is made of a 60S and a 40S subunit. There are two short rRNA molecules less than two hundred nucleotides in length (5S and 5.8S), and two RNA molecules that are much longer – one that has over five kilobases (28S), and another nearly two kilobases (18S). In all, the eukaryotic ribosome has a Svedberg coefficient of 80S. In addition, eukaryotic cells also have rRNA in mitochondria and chloroplasts. Ribosomes can be associated with the endoplasmic reticulum or be present as free floating complexes in the cytoplasm.

Functions of Ribosomal RNA

The primary function of rRNA is in protein synthesis – in binding to messenger RNA and transfer RNA to ensure that the codon sequence of the mRNA is translated accurately into amino acid sequence in proteins. To achieve this, rRNA has a distinctive three-dimensional shape involving internal loops and helices that creates specific sites within the ribosome – the A, P and E sites. The P site is for binding a growing polypeptide, the A site anchors an incoming tRNA charged with an amino acid. After peptide bond formation, the tRNA binds briefly to the E site before leaving the ribosome. In addition rRNA also has sites for binding to some ribosomal proteins and careful analysis has demarcated the exact residues in both the RNA and protein.

Ribosomal RNA are also expressed in every cell of all extant species. The sequence of the core catalytic sites are also highly conserved making rRNA an excellent tool for the study of taxonomy and phylogenetics. There is a difference in the rate of evolution of residues on the surface and interior of rRNA, and nucleotides involved in core catalytic activity, such as in the formation of a peptide bond, appear to have predated the appearance of life on earth. The extent to which two species differ in rRNA sequences can give a good estimate of their evolutionary distance.

Many antibiotics target prokaryotic rRNA and recently the binding sites for antibiotics such as streptomycin and tetracycline on rRNA have been indicated. It has also been shown that antibiotic resistance often stems from point mutations in these binding sites. For instance, the resistance of *Euglena* and *E. coli* to streptomycin stems from mutation in the 16S rRNA sequence. Similar results were found for the resistance of *Streptomyces* to Spectinomycin. Tetracycline resistance appears to come from mutations in the 30S rRNA.

In a new dimension to the function of rRNA, its precursors (preribosomal RNA) have been implicated in the generation of micro RNA that mediate inflammation and cardiac disease in response to mechanical stress. The mechanisms of this activity are still being elucidated.

Role of Ribosomal RNA in Translation

Translation of the mRNA sequence requires the involvement of rRNA at every step – initiation, elongation and termination.

Messenger RNAs carry the genetic information coded in the DNA into the cytoplasm where the nucleotide sequence is read by ribosomes in stretches of three bases called codons. Four nucleotides, Adenine, Uracil, Guanine and Cytosine, can be arranged to form a total of sixty-four triplet codons. Each codon corresponds to a single amino acid and thus codes for the protein sequence.

Prokaryotic translation begins with the 16S rRNA base pairing with the Shine-Dalgarno consensus sequence in mRNA. Since the Shine-Dalgarno sequence is 6-10 nucleotides upstream of the start codon, binding with rRNA allows the start codon to be positioned within the ribosome. This interaction is mediated by other proteins, which also recruit the larger ribosomal subunit and subsequently, the first codon is translated. In eukaryotes, eukaryotic Initiation Factors 4E and 4G (eIF4E and eIF4G) bind to the 5′ end of the mRNA, recruiting both the smaller subunit of a ribosome and a tRNA carrying methionine. The ribosome scans the mRNA to locate the start codon, after which the initiation factors dissociate from the translation machinery.

Every new amino acid, attached to a tRNA, arrives at the A site. Base pairing between the codon on the mRNA and the complementary anticodon on the tRNA changes the conformation of three residues on the 16S rRNA. These residues interact with the anticodon, stabilize the tRNA-mRNA complex and enzymatic activity of the rRNA positions the aminoacyl-tRNA fully within the A site.

The polypeptide that has been synthesized so far is bound to the P site on the ribosome. Ribosomal RNA in the larger subunit catalyzes the reaction that forms a peptide bond between the amino acid in the A site and the growing polypeptide chain in the P site. Polypeptide synthesis is terminated when the ribosome reaches a stop codon, and rRNA catalyzes the addition of a water molecule to the polypeptide in the P site.

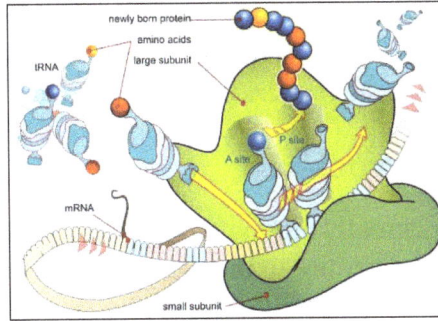

Ribosome mRNA translation.

Messenger RNA

Messenger RNA (mRNA) is a large family of RNA molecules that convey genetic information from DNA to the ribosome, where they specify the amino acid sequence of the protein products of gene expression. RNA polymerase transcribes primary transcript mRNA (known as pre-mRNA) into processed, mature mRNA. This mature mRNA is then translated into a polymer of amino acids: a protein, as summarized in the central dogma of molecular biology.

As in DNA, mRNA genetic information is in the sequence of nucleotides, which are arranged into codons consisting of three base pairs each. Each codon encodes for a specific amino acid, except the stop codons, which terminate protein synthesis. This process of translation of codons into amino acids requires two other types of RNA: Transfer RNA (tRNA), that mediates recognition of the codon and provides the corresponding amino acid, and ribosomal RNA (rRNA), that is the central component of the ribosome's protein-manufacturing machinery.

It should not be confused with mitochondrial DNA.

Synthesis, Processing and Function

The brief existence of an mRNA molecule begins with transcription, and ultimately ends in degradation. During its life, an mRNA molecule may also be processed, edited, and transported prior to translation. Eukaryotic mRNA molecules often require extensive processing and transport, while prokaryotic mRNA molecules do not. A molecule of eukaryotic mRNA and the proteins surrounding it are together called a messenger RNP.

Transcription

Transcription is when RNA is made from DNA. During transcription, RNA polymerase makes a copy of a gene from the DNA to mRNA as needed. This process is similar in eukaryotes and prokaryotes. One notable difference, however, is that eukaryotic RNA polymerase associates with mRNA-processing enzymes during transcription so that processing can proceed quickly after the start of transcription. The short-lived, unprocessed or partially processed product is termed *precursor mRNA*, or *pre-mRNA*; once completely processed, it is termed *mature mRNA*.

Eukaryotic Pre-mRNA Processing

Processing of mRNA differs greatly among eukaryotes, bacteria, and archea. Non-eukaryotic mRNA is, in essence, mature upon transcription and requires no processing, except in rare cases. Eukaryotic pre-mRNA, however, requires extensive processing.

5' Cap Addition

A *5' cap* (also termed an RNA cap, an RNA 7-methylguanosine cap, or an RNA m7G cap) is a modified guanine nucleotide that has been added to the "front" or 5' end of a eukaryotic messenger RNA shortly after the start of transcription. The 5' cap consists of a terminal 7-methylguanosine residue that is linked through a 5'-5'-triphosphate bond to the first transcribed nucleotide. Its presence is critical for recognition by the ribosome and protection from RNases.

Cap addition is coupled to transcription, and occurs co-transcriptionally, such that each influences the other. Shortly after the start of transcription, the 5' end of the mRNA being synthesized is bound by a cap-synthesizing complex associated with RNA polymerase. This enzymatic complex catalyzes the chemical reactions that are required for mRNA capping. Synthesis proceeds as a multi-step biochemical reaction.

Editing

In some instances, an mRNA will be edited, changing the nucleotide composition of that mRNA. An example in humans is the apolipoprotein B mRNA, which is edited in some tissues, but not others. The editing creates an early stop codon, which, upon translation, produces a shorter protein.

Polyadenylation

Polyadenylation is the covalent linkage of a polyadenylyl moiety to a messenger RNA molecule. In eukaryotic organisms most messenger RNA (mRNA) molecules are polyadenylated at the 3' end, but recent studies have shown that short stretches of uridine (oligouridylation) are also common. The poly(A) tail and the protein bound to it aid in protecting mRNA from degradation by exonucleases. Polyadenylation is also important for transcription termination, export of the mRNA from the nucleus, and translation. mRNA can also be polyadenylated in prokaryotic organisms, where poly(A) tails act to facilitate, rather than impede, exonucleolytic degradation.

Polyadenylation occurs during and/or immediately after transcription of DNA into RNA. After transcription has been terminated, the mRNA chain is cleaved through the action of an endonuclease complex associated with RNA polymerase. After the mRNA has been cleaved, around 250 adenosine residues are added to the free 3' end at the cleavage site. This reaction is catalyzed by polyadenylate polymerase. Just as in alternative splicing, there can be more than one polyadenylation variant of an mRNA.

Polyadenylation site mutations also occur. The primary RNA transcript of a gene is cleaved at the poly-A addition site, and 100–200 A's are added to the 3' end of the RNA. If this site is altered, an abnormally long and unstable mRNA construct will be formed.

Transport

Another difference between eukaryotes and prokaryotes is mRNA transport. Because eukaryotic transcription and translation is compartmentally separated, eukaryotic mRNAs must be exported from the nucleus to the cytoplasm—a process that may be regulated by different signaling pathways. Mature mRNAs are recognized by their processed modifications and then exported through the nuclear pore by binding to the cap-binding proteins CBP20 and CBP80, as well as the transcription/export complex (TREX). Multiple mRNA export pathways have been identified in eukaryotes.

In spatially complex cells, some mRNAs are transported to particular subcellar destinations. In mature neurons, certain mRNA are transported from the soma to dendrites. One site of mRNA translation is at polyribosomes selectively localized beneath synapses. The mRNA for Arc/Arg3.1 is induced by synaptic activity and localizes selectively near active synapses based on signals generated by NMDA receptors. Other mRNAs also move into dendrites in response to external stimuli, such as β-actin mRNA. Upon export from the nucleus, actin mRNA associates with ZBP1 and the 40S subunit. The complex is bound by a motor protein and is transported to the target location (neurite extension) along the cytoskeleton. Eventually ZBP1 is phosphorylated by Src in order for translation to be initiated. In developing neurons, mRNAs are also transported into growing axons and especially growth cones. Many mRNAs are marked with so-called "zip codes," which target their transport to a specific location.

Translation

Because prokaryotic mRNA does not need to be processed or transported, translation by the ribosome can begin immediately after the end of transcription. Therefore, it can be said that prokaryotic translation is *coupled* to transcription and occurs *co-transcriptionally.*

Eukaryotic mRNA that has been processed and transported to the cytoplasm (i.e., mature mRNA) can then be translated by the ribosome. Translation may occur at ribosomes free-floating in the cytoplasm, or directed to the endoplasmic reticulum by the signal recognition particle. Therefore, unlike in prokaryotes, eukaryotic translation *is not* directly coupled to transcription. It is even possible in some contexts that reduced mRNA levels are accompanied by increased protein levels, as has been observed for mRNA/protein levels of EEF1A1 in breast cancer.

Structure

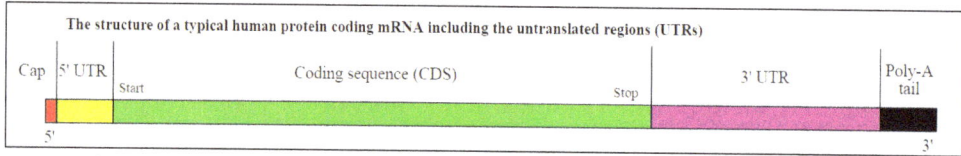

The structure of a mature eukaryotic mRNA. A fully processed mRNA includes a 5' cap, 5' UTR, coding region, 3' UTR, and poly(A) tail.

Coding Regions

Coding regions are composed of codons, which are decoded and translated (in eukaryotes usually into one and in prokaryotes usually into several) into proteins by the ribosome. Coding regions begin with the start codon and end with a stop codon. In general, the start codon is an AUG triplet and the stop codon is UAA, UAG, or UGA. The coding regions tend to be stabilised by internal base pairs, this impedes degradation. In addition to being protein-coding, portions of coding regions may serve as regulatory sequences in the pre-mRNA as exonic splicing enhancers or exonic splicing silencers.

Untranslated Regions

Untranslated regions (UTRs) are sections of the mRNA before the start codon and after the stop codon that are not translated, termed the five prime untranslated region (5' UTR) and three prime untranslated region (3' UTR), respectively. These regions are transcribed with the coding region and thus are exonic as they are present in the mature mRNA. Several roles in gene expression have been attributed to the untranslated regions, including mRNA stability, mRNA localization, and translational efficiency. The ability of a UTR to perform these functions depends on the sequence of the UTR and can differ between mRNAs. Genetic variants in 3' UTR have also been implicated in disease susceptibility because of the change in RNA structure and protein translation.

The stability of mRNAs may be controlled by the 5' UTR and/or 3' UTR due to varying affinity for RNA degrading enzymes called ribonucleases and for ancillary proteins that can promote or inhibit RNA degradation.

Translational efficiency, including sometimes the complete inhibition of translation, can be controlled by UTRs. Proteins that bind to either the 3' or 5' UTR may affect translation by influencing the ribosome's ability to bind to the mRNA. MicroRNAs bound to the 3' UTR also may affect translational efficiency or mRNA stability.

Cytoplasmic localization of mRNA is thought to be a function of the 3' UTR. Proteins that are needed in a particular region of the cell can also be translated there; in such a case, the 3' UTR may contain sequences that allow the transcript to be localized to this region for translation.

Some of the elements contained in untranslated regions form a characteristic secondary structure when transcribed into RNA. These structural mRNA elements are involved in regulating the mRNA. Some, such as the SECIS element, are targets for proteins to bind. One class of mRNA element, the riboswitches, directly bind small molecules, changing their fold to modify levels of transcription or translation. In these cases, the mRNA regulates itself.

Poly(A) Tail

The 3' poly(A) tail is a long sequence of adenine nucleotides (often several hundred) added to the 3' end of the pre-mRNA. This tail promotes export from the nucleus and translation, and protects the mRNA from degradation.

Monocistronic Versus Polycistronic mRNA

An mRNA molecule is said to be monocistronic when it contains the genetic information to translate only a single protein chain (polypeptide). This is the case for most of the eukaryotic mRNAs. On the other hand, polycistronic mRNA carries several open reading frames (ORFs), each of which is translated into a polypeptide. These polypeptides usually have a related function (they often are the subunits composing a final complex protein) and their coding sequence is grouped and regulated together in a regulatory region, containing a promoter and an operator. Most of the mRNA found in bacteria and archaea is polycistronic, as is the human mitochondrial genome. Dicistronic or bicistronic mRNA encodes only two proteins.

mRNA Circularization

In eukaryotes mRNA molecules form circular structures due to an interaction between the eIF4E and poly(A)-binding protein, which both bind to eIF4G, forming an mRNA-protein-mRNA bridge. Circularization is thought to promote cycling of ribosomes on the mRNA leading to time-efficient translation, and may also function to ensure only intact mRNA are translated (partially degraded mRNA characteristically have no m7G cap, or no poly-A tail).

Other mechanisms for circularization exist, particularly in virus mRNA. Poliovirus mRNA uses a cloverleaf section towards its 5' end to bind PCBP2, which binds poly(A)-binding protein, forming the familiar mRNA-protein-mRNA circle. Barley yellow dwarf virus has binding between mRNA segments on its 5' end and 3' end (called kissing stem loops), circularizing the mRNA without any proteins involved.

RNA virus genomes (the + strands of which are translated as mRNA) are also commonly circularized. During genome replication the circularization acts to enhance genome replication speeds, cycling viral RNA-dependent RNA polymerase much the same as the ribosome is hypothesized to cycle.

Degradation

Different mRNAs within the same cell have distinct lifetimes (stabilities). In bacterial cells, individual mRNAs can survive from seconds to more than an hour. However, the lifetime averages between 1 and 3 minutes, making bacterial mRNA much less stable than eukaryotic mRNA. In mammalian cells, mRNA lifetimes range from several minutes to days. The greater the stability of an mRNA the more protein may be produced from that mRNA. The limited lifetime of mRNA enables a cell to alter protein synthesis rapidly in response to its changing needs. There are many mechanisms that lead to the destruction of an mRNA, some of which are described below.

Prokaryotic mRNA Degradation

In general, in prokaryotes the lifetime of mRNA is much shorter than in eukaryotes. Prokaryotes degrade messages by using a combination of ribonucleases, including endonucleases, 3' exonucleases, and 5' exonucleases. In some instances, small RNA molecules (sRNA) tens to hundreds of nucleotides long can stimulate the degradation of specific mRNAs by base-pairing with complementary sequences and facilitating ribonuclease cleavage by RNase III. It was recently shown that bacteria also have a sort of 5' cap consisting of a triphosphate on the 5' end. Removal of two of the phosphates leaves a 5' monophosphate, causing the message to be destroyed by the exonuclease RNase J, which degrades 5' to 3'.

Eukaryotic mRNA Turnover

Inside eukaryotic cells, there is a balance between the processes of translation and mRNA decay. Messages that are being actively translated are bound by ribosomes, the eukaryotic initiation factors eIF-4E and eIF-4G, and poly(A)-binding protein. eIF-4E and eIF-4G block the decapping enzyme (DCP2), and poly(A)-binding protein blocks the exosome complex, protecting the ends of the message. The balance between translation and decay is reflected in the size and abundance of cytoplasmic structures known as P-bodies The poly(A) tail of the mRNA is shortened by specialized exonucleases that are targeted to specific messenger RNAs by a combination of cis-regulatory sequences on the RNA and trans-acting RNA-binding proteins. Poly(A) tail removal is thought to disrupt the circular structure of the message and destabilize the cap binding complex. The message is then subject to degradation by either the exosome complex or the decapping complex. In this way, translationally inactive messages can be destroyed quickly, while active messages remain intact. The mechanism by which translation stops and the message is handed-off to decay complexes is not understood in detail.

AU-rich Element Decay

The presence of AU-rich elements in some mammalian mRNAs tends to destabilize those transcripts through the action of cellular proteins that bind these sequences and stimulate poly(A) tail removal. Loss of the poly(A) tail is thought to promote mRNA degradation by facilitating attack by both the exosome complex and the decapping complex. Rapid mRNA degradation via AU-rich elements is a critical mechanism for preventing the overproduction of potent cytokines such as tumor necrosis factor (TNF) and granulocyte-macrophage colony stimulating factor (GM-CSF). AU-rich elements also regulate the biosynthesis of proto-oncogenic transcription factors like c-Jun and c-Fos.

Nonsense Mediated Decay

Eukaryotic messages are subject to surveillance by nonsense mediated decay (NMD), which checks for the presence of premature stop codons (nonsense codons) in the message. These can arise via incomplete splicing, V(D)J recombination in the adaptive immune system, mutations in DNA, transcription errors, leaky scanning by the ribosome causing a frame shift, and other causes. Detection of a premature stop codon triggers mRNA degradation by 5' decapping, 3' poly(A) tail removal, or endonucleolytic cleavage.

Small Interfering RNA (siRNA)

In metazoans, small interfering RNAs (siRNAs) processed by Dicer are incorporated into a complex known as the RNA-induced silencing complex or RISC. This complex contains an endonuclease that cleaves perfectly complementary messages to which the siRNA binds. The resulting mRNA fragments are then destroyed by exonucleases. siRNA is commonly used in laboratories to block the function of genes in cell culture. It is thought to be part of the innate immune system as a defense against double-stranded RNA viruses.

MicroRNA (miRNA)

MicroRNAs (miRNAs) are small RNAs that typically are partially complementary to sequences in metazoan messenger RNAs. Binding of a miRNA to a message can repress translation of that message and accelerate poly(A) tail removal, thereby hastening mRNA degradation. The mechanism of action of miRNAs is the subject of active research.

Other Decay Mechanisms

There are other ways by which messages can be degraded, including non-stop decay and silencing by Piwi-interacting RNA (piRNA), among others.

mRNA-based Therapeutics

Full length mRNA molecules have been proposed as therapeutics since the beginning of the biotech era but there was little traction until the 2010s, when Moderna Therapeutics was founded and managed to raise almost a billion dollars in venture funding in its first three years.

Theoretically, the administered mRNA sequence can cause a cell to make a protein, which in turn could directly treat a disease or could function as a vaccine; more indirectly the protein could drive an endogenous stem cell to differentiate in a desired way.

The primary challenges of RNA therapy center on delivering the RNA to directed cells, more even than determining what sequence to deliver. Naked RNA sequences will naturally degrade after preparation; they may trigger the body's immune system to attack them as an invader; and they are impermeable to the cell membrane. Once within the cell, they must then leave the cell's transport mechanism to take action within the cytoplasm, which houses the ribosomes that direct manufacture of proteins.

Transfer RNA

Transfer RNAs or tRNAs are molecules that act as temporary carriers of amino acids, bringing the appropriate amino acids to the ribosome based on the messenger RNA (mRNA) nucleotide sequence. In this way, they act as the intermediaries between nucleotide and amino acid sequences.

tRNAs are ribonucleic acids and therefore capable of forming hydrogen bonds with mRNA. Additionally, they can also form ester linkages with amino acids, and therefore, can physically bring mRNA and amino acids together during the process of translation. They pair with mRNA in a complementary and antiparallel manner, and each tRNA can base pair with a stretch of three nucleotides on mRNA. These sets of three nucleotides on the mRNA are called codons and the corresponding sequence on the tRNA is called the anticodon. Base pairing between the codon and anticodon brings specificity to the process of translation. On one end of the tRNA, an appropriate amino acid is attached to its 3' hydroxyl group based on the anticodon and the ribosome catalyzes the formation of a peptide bond between this amino acid and the elongating polypeptide chain.

tRNA Structure and Function

Transfer RNAs are coded by a number of genes, and are usually short molecules, between 70-90 nucleotides (5 nm) in length. The two most important parts of a tRNA are its anticodon and the terminal 3' hydroxyl group, which can form an ester linkage with an amino acid. However, there are other aspects to a tRNA's structure such as the D-arm and T-arm, which contribute to its high level of specificity and efficiency. Only 1 in 10,000 amino acids are incorrectly attached to a tRNA, which is a remarkable number given the chemical similarities between many amino acids.

Transfer RNAs have a sugar-phosphate backbone like all other cellular nucleic acids and the orientation of the ribose sugar gives rise to directionality in the molecule. One end of the RNA has a reactive phosphate group attached to the fifth carbon atom of ribose while the other end has a free hydroxyl group on the third carbon atom. This gives rise to the 5' and 3' ends of the RNA since all the other phosphate and hydroxyl groups are involved in phosphodiester bonds within the nucleic acid.

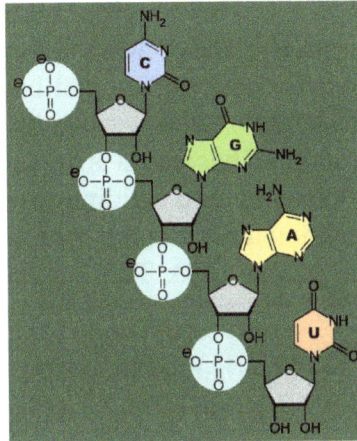
RNA-Nucleobases.

The last three bases on the 3' end of tRNA are always CCA – two cytosines followed by one adenine base. This stretch is part of the acceptor arm of the molecule, where an amino acid is covalently attached to the hydroxyl group on the ribose sugar of the terminal adenine nucleotide. The acceptor arm also contains parts of the 5' end of the tRNA, with a stretch of 7-9 nucleotides from opposite ends of the molecule base pairing with each other.

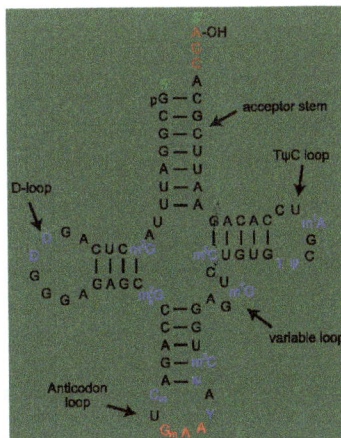
TRNA-Phe from yeast.

The anticodon loop, which pairs with mRNA, determines which amino acid is attached to the acceptor stem. The anticodon loop is recognized by aminoacyl tRNA synthetase (AATS), the enzyme that chemically links a tRNA to an amino acid through a high-energy bond. AATS 'reads' the anticodon and also recognizes the D-arm located downstream from the 5' end of the tRNA.

The D-arm is made of a double-stranded stem region formed by internal base pairing as well as a loop structure of unpaired nucleotides. The D-arm is a highly variable region and plays an important role in stabilizing the RNA's tertiary structure and also influences the kinetics and accuracy of translation at the ribosome.

The other structure that influences the role of tRNA in translation is the T-arm. Similar to the D-arm, it contains a stretch of nucleotides that base pair with each other and a loop that is single stranded. The paired region is called the 'stem' and mostly contains 5 base pairs. The loop contains modified bases and is also called the TΨC arm, to specify the presence thymidine, pseudouridine and cytidine residues (modified bases). tRNA molecules are unusual in containing a high number of modified bases as well as containing thymidine, usually seen only in DNA. The T-arm is involved in the interaction of tRNA with the ribosome.

Finally, a variable arm containing less than 20 nucleotides is situated between the anticodon loop and the T-arm. It plays a role in AATS recognition of tRNA, but could be absent in some species.

The secondary structure of tRNA containing the acceptor region, D- and T-arms and the anticodon loop is said to resemble a cloverleaf. After the RNA folds into its tertiary structure, it is L-shaped, with the acceptor stem and T-arm forming an extended helix and the anticodon loop and D-arm similarly making another extended helix. These two helices align perpendicularly to each other in a way that brings the D-arm and T-arm into close proximity while the anticodon loop and the acceptor arm are positioned on opposite ends of the molecule.

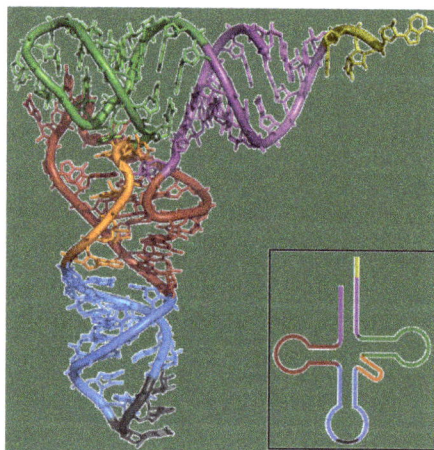

In this image, the 3' CCA region is in yellow, the acceptor arm is in purple, the variable loop in orange, the D-arm is in red, the T-arm in green and the anticodon loop is in blue.

Types of tRNA

A tRNA can be classified based on the amino acid it carries, giving rise to 20 different tRNAs. Alternatively, they can also be grouped based on their anticodon. There are 64 possible codons arising from a combination of four nucleotides. Of these, 3 are stop codons that signal the end of translation. This gives rise to a situation where one amino acid is represented by multiple codons and the AATS, as well as the tRNAs have to accommodate this redundancy. However, very few species have exactly 61 tRNAs, which gives rise to the question of how every codon is recognized by a specific tRNA. In many species, the number far exceeds 61 and different tRNAs carrying the same anticodon could display varying efficiency in translation, adding a layer of regulation to the process of protein synthesis.

tRNAs interact with codons on the mRNA through their anticodon loop. Base pairing between the codon and anticodon ensures specificity during translation. However, the first base of the anticodon, that pairs with the 'wobble' or third position in a codon is often modified to allow the tRNA to hydrogen bond with three, instead of one base. Thus a single tRNA has the option of recognizing and base pairing with three codons, which code for the same amino acid. There are 20 AATS, one for each amino acid. This group of enzymes can recognize all the anticodons representing a particular amino acid and therefore act as the second arm of the machinery that handles genetic code redundancy.

Finally, these molecules can also be classified into three categories – those carrying canonical amino acids attached to the correct tRNA, those that are incorrectly attached, and those carrying modified amino acids such as selenocysteine for non-canonical elongation.

Post-Transcriptional Modification of tRNA

There are nearly 500 genes coding for tRNAs in the human genome, and 300 gene fragments associated with these RNA. These genes are transcribed by RNA polymerase III and the transcript undergoes extensive modification, especially in eukaryotes. Introns are spliced, the intron-exon boundary is acted on by endonucleases, the 5' and 3' ends of the RNA are processed and enzymes add the terminal CCA residues to the 3' end of the tRNA. The CCA residues could become aminoacylated in the nucleus itself and this charged tRNA could then be exported from the nucleus.

Additionally, many bases on the tRNA are also modified, especially by methylation (addition of a methyl group) and deamidation (removal of an amide group). Particularly, the first base of the anticodon that pairs with the 'wobble' position on the codon is modified to allow unusual types of base pairing. Adenine can be modified to form inosine, which expands the pairing possibilities to include uracil, cytosine and adenine. Pseudouridine is another common modified base, derived from uridine residues through enzyme-mediated isomerization. It is said to play a role in the structural integrity of the tRNA molecule, being involved in stiffening the nearby sugar-phosphate backbone and also influencing base stacking of proximal regions. Lysidine is an unusual base formed when a lysine amino acid is attached to cytidine residue. Lysidine pairs specifically with adenosine, a property that is used by the isoleucine tRNA to ensure translation specificity.

AATS attach the appropriate amino acid to tRNA molecules based on their anticodon. These enzymes contain binding sites for the amino acid, tRNA as well as ATP and hydrolyze ATP to AMP and attach the amino acid to the ribose sugar of the last nucleotide on tRNA. The tRNA is now considered 'charged' and can participate in the protein synthesizing reactions on the ribosome. This reaction often occurs in the cytoplasm, though it has also been observed in the nucleus.

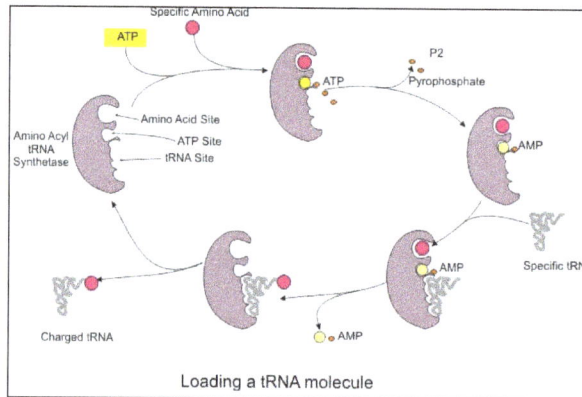

Charge tRNA.

The enzyme binds to many regions of the tRNA to ensure high specificity in the reaction and even proofreads its own reaction since many amino acids have similar structures.

Mature tRNA then binds specific export factors that export it from the nucleus, using the RanGTP system. The acceptor arm and T-arm play an important role in this process, and there is extensive interaction between the export factors and the RNA molecule, allowing only fully processed, complete tRNAs to move to the cytoplasm.

tRNA Interaction with Ribosome

The ribosome contains three important regions – the P (peptidyl) site containing the growing polypeptide, the A (acceptor) site that receives a new charged tRNA and the E (exit) site through which a deacylated tRNA leaves the ribosome. These sites span both the subunits of the ribosome and are denoted as P/P or A/A sites with the first letter referring to the site on the smaller subunit. For instance, the P/P site binds to tRNA anchoring a polypeptide chain while the A/A site anchors an incoming charged tRNA. The peptidyl-tRNA on the P/P site transfers the growing polypeptide to the tRNA on the A/A site and undergoes deacylation. To continue the process of translation, the ribosome moves ahead by one codon, making the tRNA on the P/P site shift to a transient P/E configuration and then to the E/E site before leaving the ribosome. Similarly, the tRNA on the A/A site adopts a temporary A/P binding conformation before settling at the P/P site, ready for the next amino acid to continue translation.

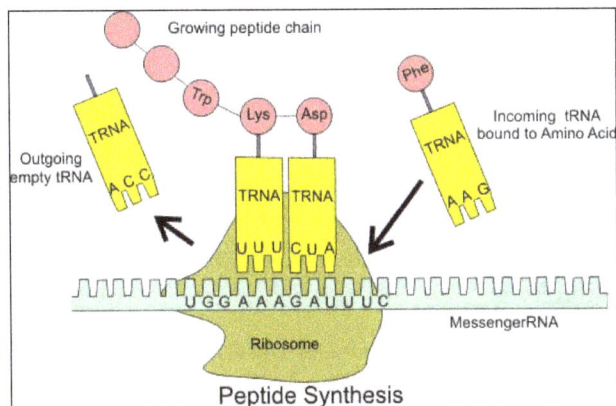

Peptide Synthesis

RNA Transcription

Transcription refers to the first step of gene expression where an RNA polymer is created from a DNA template. This reaction is catalyzed by enzymes called RNA polymerases and the RNA polymer is antiparallel and complementary to the DNA template. The stretch of DNA that codes for an RNA transcript is called a transcription unit and could contain more than one gene.

These RNA transcripts can either be used as messengers to drive the synthesis of proteins or be involved in a number of different cellular processes. These functional or non-coding RNA could be transfer RNA (tRNA), ribosomal RNA (rRNA), or direct gene regulation through RNA interference (RNAi) and the formation of heterochromatin.

Function of Transcription

Life on earth is said to have begun from self-replicating RNA since it is the only class of molecules capable of both catalysis and carrying genetic information. With evolution, proteins took over catalysis because they are capable of a greater variety of sequences and structures. Additionally, the bonds on the sugar phosphate backbone of RNA are vulnerable to even mild changes in pH and can undergo alkaline hydrolysis. Therefore, DNA emerged as the preferred molecule for carrying genetic information since it is more stable and resistant to degradation. Transcription maintains the link between these two molecules and allows cells to use a stable nucleic acid as the genetic material while retaining most of their protein synthesis machinery.

In addition, separating DNA from the site for protein synthesis also protects genetic material from the biochemical and biophysical stresses of complex, multilayered processes. Small errors in the RNA transcript can be overcome since the RNA molecule has a short half-life, but changes to the DNA become heritable mutations. In addition, transcription adds another layer for intricate gene regulation, especially in species with large genomes that require minute adjustments in metabolism.

In eukaryotes, transcription also plays an important role in transferring the information from DNA to the cytoplasm because the nuclear pore is too small to allow ribosomes, proteins or chromosomes to pass through. While the nuclear pore has a diameter of about 5-10 nm, ribosomes are between 25-30 nm in size, many proteins are wider than 10 nm and fully condensed chromosomes can be over 2000 nm in size. Therefore, the primary machinery for protein synthesis cannot enter the nucleus and stretches of DNA cannot exit the nucleus.

Mechanism of Transcription

Transcription creates a single stranded RNA molecule from double stranded DNA. Therefore, only the information in one of the strands is transferred into the nucleotide sequence of RNA. One strand of DNA is called the coding strand and the other is the template strand. Transcription machinery interacts with the template strand to produce an mRNA whose sequence resembles the coding strand. Other names for the template strand include antisense strand and master strand.

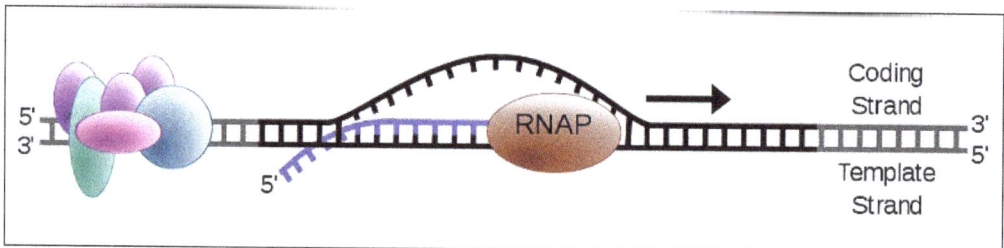

Two different genes on the same DNA molecule can have coding sequences on different strands.

Transcriptional activity is particularly high in the G1 and G2 phases of the cell cycle when the cell is either recovering from mitosis or preparing for the dramatic events of the next cycle of cell division.

Transcription Initiation

Transcription begins with the binding of an RNAP in the presence of general transcription factors to the promoter region upstream of the transcription start site on the DNA. Prokaryotic RNAP binds with a sigma factor, while eukaryotic RNA polymerases can interact with a number of transcription factors as well as activator and repressor proteins. Initially, after the binding of RNAP to the promoter region, the DNA remains in a double-stranded form. This is called a 'closed complex' between DNA and RNAP. Thereafter, RNAP along with transcription factors unwinds a segment of the DNA and interacts with the exposed nucleotides in an open complex creating a 'transcription bubble'. RNAP then cruises along the DNA scanning for the transcription start site inside the bubble. Once the start site is located, the first two nucleotides of the transcript are bonded to each other.

Escape from Promoter

After the first few nucleotides are added to the putative RNA transcript, RNAP enters a critical, unstable phase. It can either continue towards productive initiation, or pull DNA towards itself, creating scrunched open DNA inside the polymerase. If RNAP rewinds the downstream portion of the DNA, the putative RNA transcript is released because the DNA-RNAP complex reverts to its initial open configuration. This is called abortive initiation.

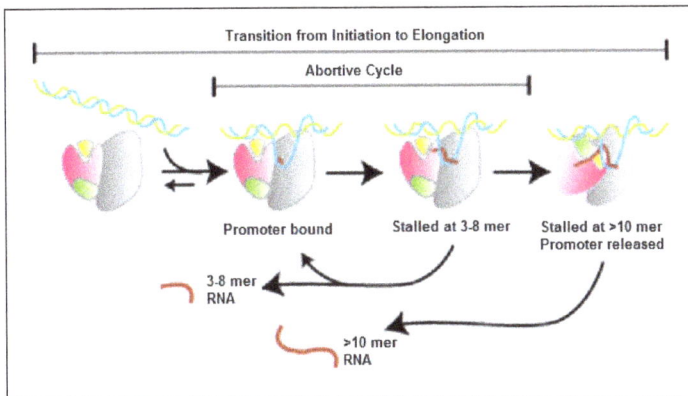

However, if the upstream portion of DNA is rewound and ejected from the enzyme, RNAP moves ahead. Its interaction with the promoter region is broken and the RNA transcript reaches a length of 14-15 nucleotides. This is called escape from the promoter and is accompanied by changes to protein-protein and protein-DNA interactions. Some transcription factors are released and transcription moves towards the elongation phase.

Transcription Elongation

Once a short RNA oligonucleotide of more than 15 bases is formed, RNAP proceeds along the template DNA strand. The transcript is identical to the coding strand, except that the nucleotide backbone has ribose sugar instead of deoxyribose, and adenine base pairs with uracil, instead of thymine. RNAP can catalyze the formation of a phosphodiester bond between the fifth carbon atom of an incoming nucleotide and the third carbon atom of the last nucleotide in the existing transcript.

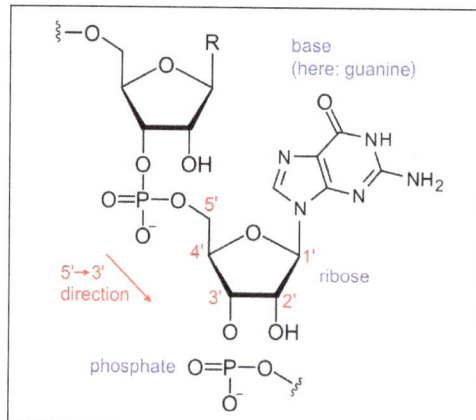

Since the RNA molecule has a free phosphate attached to the fifth carbon on the first nucleotide and a free hydroxyl group on the third carbon of the last nucleotide, RNA is said to be transcribed in a 5' to 3' direction.

Transcription Termination

Unlike DNA replication, where the DNA polymerase continues to add nucleotides till it reaches the end of the molecule, transcription has to be terminated at a particular location for effective gene regulation and expression. Prokaryotic transcription termination can occur through the formation of a double-stranded region within the RNA or through the action of a protein called Rho.

The first method involves the transcription of a G:C rich region followed by a string of uracils that form weak hydrogen bonds with template DNA. The G:C rich region can loop over itself to form a hairpin-like structure stalling the RNAP and transcription machinery. This, combined with the weaker bonds between uracil and the template DNA can prise the RNA away from the transcription machinery and lead to termination. This process also involves a protein called NusA.

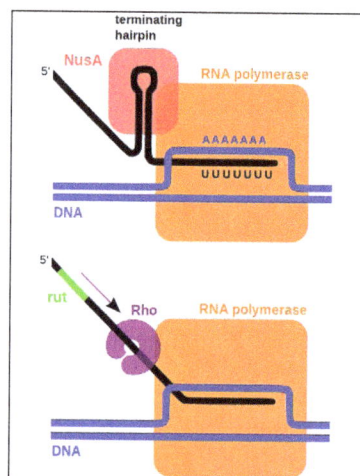

Rho-dependent transcription termination involves the binding of Rho protein to a sequence on the transcribed RNA. This sequence, which is downstream from translation stop codons, allows Rho to bind to RNA and cruise along the transcript in an ATP-dependent manner. When it encounters a stalled RNAP, it binds to the enzyme and causes the transcript and its associated machinery to dissociate from the DNA.

Eukaryotic transcription termination is much less understood, and most of the work has focussed on the mechanisms of RNAP II. Transcription termination in eukaryotes is also coupled with post-transcriptional modifications and processing before the mature RNA is exported to the cytoplasm.

Types of RNA Transcripts

Traditionally, three types of RNA transcripts were known – messenger RNA (mRNA), tRNA and rRNA – and all three are intimately associated with protein synthesis. While mRNA determines amino acid sequence, tRNA and rRNA are crucial for the mechanism of translating the mRNA code.

mRNA polymerization from DNA containing protein coding genes is catalyzed by RNA polymerase II. Occasionally, proteins that are used together are coded as a single unit, in one long mRNA molecule and this is particularly common among prokaryotes. DNA sequences upstream of the coding sequence contain docking sites for the transcription machinery as well as regulatory factors that modulate the timing and quantity of transcriptional activity. mRNA is then modified and processed to give rise to the final transcript used for translation.

rRNA constitutes nearly fifty percent of the RNA of a cell and is transcribed by RNA polymerase I in specialized regions of the nucleus called the nucleolus. Nucleoli appear as dense spherical structures around the loci that code for rRNA. Prokaryotic rRNA is of three types and eukaryotic ribosomes are made of four types of rRNA with the largest one containing over 5000 nucleotides. These RNA molecules determine the three-dimensional structure of ribosomes.

RNA polymerase III catalyzes the transcription of tRNA precursors in the nucleus. Promoter sequences controlling the expression of tRNA genes can be intragenic, located inside the coding sequence of the gene. tRNA precursors undergo extensive modifications including splicing. Prokaryotic tRNAs retain their catalytic activity and can self-splice, whereas eukaryotic post-transcriptional modification is carried out by special endonuclease enzymes. These endonucleases recognize specific structural motifs within the tRNA that target the sequence for splicing.

In addition to these three types of RNA, the cell contains a number of smaller RNA involved in various cellular activities. These include gene regulation (mediated by micro RNA and sequences in the 5′ untranslated regions of mRNA transcripts), post-transcriptional modification (small nuclear RNA, small nucleolar RNA), genome defense

(Piwi-interacting RNA and CRISPR) and the maintenance of genomic structure (telomeres and RNA transcripts that silence X-chromosomes).

Differences between Prokaryotic and Eukaryotic Transcription

The obvious difference between prokaryotic and eukaryotic transcription is the presence of a nuclear membrane in eukaryotes. Eukaryotic RNA transcripts need to be exported from the nucleus, whereas prokaryotes conduct coupled transcription and translation in the cytoplasm. This is possible because the prokaryotic transcript does not undergo extensive modification and prokaryotes do not need transcription factors for initiation. Therefore, the transcription machinery is simpler and can simultaneously accommodate the enzymes of translation.

Prokaryotes also have only one RNA polymerase to catalyze all the transcription reactions of the cell and a single RNA transcript can direct the synthesis of multiple proteins. These mRNA are called polycistronic mRNA. Often, all the genes involved in one biochemical pathway are transcribed and translated together, allowing the entire pathway to be regulated as a single unit. In eukaryotes, polycistronic mRNA can be found in chloroplasts.

RNA Processing

Eukaryotic pre-mRNA receives a 5′ cap and a 3′ poly (A) tail before introns are removed and the mRNA is considered ready for translation.

Pre-mRNA Processing

The eukaryotic pre-mRNA undergoes extensive processing before it is ready to be translated. The additional steps involved in eukaryotic mRNA maturation create a molecule with a much longer half-life than a prokaryotic mRNA. Eukaryotic mRNAs last for several hours, whereas the typical E. coli mRNA lasts no more than five seconds.

Pre-mRNAs are first coated in RNA-stabilizing proteins; these protect the pre-mRNA from degradation while it is processed and exported out of the nucleus. The three most important steps of pre-mRNA processing are the addition of stabilizing and signaling factors at the 5′ and 3′ ends of the molecule, and the removal of intervening sequences that do not specify the appropriate amino acids. In rare cases, the mRNA transcript can be "edited" after it is transcribed.

5′ Capping

While the pre-mRNA is still being synthesized, a 7-methylguanosine cap is added to the 5′ end of the growing transcript by a 5′-to-5′ phosphate linkage. This moiety protects the nascent mRNA from degradation. In addition, initiation factors involved in protein synthesis recognize the cap to help initiate translation by ribosomes.

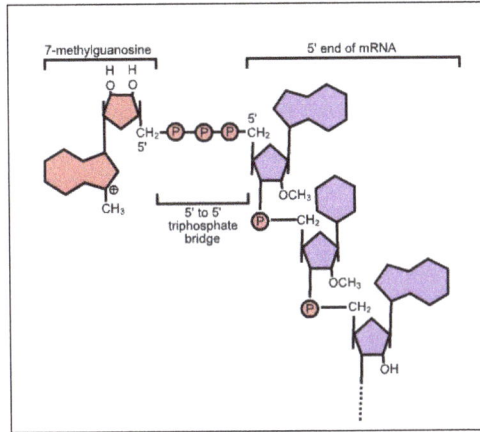

5′ cap structure: Capping of the pre-mRNA involves the addition of 7-methylguanosine (m7G) to the 5′ end. The cap protects the 5′ end of the primary RNA transcript from attack by ribonucleases and is recognized by eukaryotic initiation factors involved in assembling the ribosome on the mature mRNA prior to initiating translation.

3′ Poly-A Tail

While RNA Polymerase II is still transcribing downstream of the proper end of a gene, the pre-mRNA is cleaved by an endonuclease-containing protein complex between an AAUAAA consensus sequence and a GU-rich sequence. This releases the functional pre-mRNA from the rest of the transcript, which is still attached to the RNA Polymerase. An enzyme called poly (A) polymerase (PAP) is part of the same protein complex that cleaves the pre-mRNA and it immediately adds a string of approximately 200 A nucleotides, called the poly (A) tail, to the 3′ end of the just-cleaved pre-mRNA. The poly (A) tail protects the mRNA from degradation, aids in the export of the mature mRNA to the cytoplasm, and is involved in binding proteins involved in initiating translation.

Pre-mRNA Splicing

Eukaryotic genes are composed of exons, which correspond to protein-coding sequences (ex-on signifies that they are expressed), and intervening sequences called introns (int-ron denotes their intervening role), which may be involved in gene regulation, but are removed from the pre-mRNA during processing. Intron sequences in mRNA do not encode functional proteins.

Discovery of Introns

The discovery of introns came as a surprise to researchers in the 1970s who expected that pre-mRNAs would specify protein sequences without further processing, as they had observed in prokaryotes. The genes of higher eukaryotes very often contain one or more introns. While these regions may correspond to regulatory sequences, the biological significance of having many introns or having very long introns in a gene is unclear. It is possible that introns slow down gene expression because it takes longer to transcribe pre-mRNAs with lots of introns. Alternatively, introns may be nonfunctional sequence remnants left over from the fusion of ancient genes throughout evolution. This is supported by the fact that separate exons often encode separate protein subunits or domains. For the most part, the sequences of introns can be mutated without ultimately affecting the protein product.

Intron Processing

All introns in a pre-mRNA must be completely and precisely removed before protein synthesis. If the process errs by even a single nucleotide, the reading frame of the rejoined exons would shift, and the resulting protein would be dysfunctional. The process of removing introns and reconnecting exons is called splicing. Introns are removed and degraded while the pre-mRNA is still in the nucleus. Splicing occurs by a sequence-specific mechanism that ensures introns will be removed and exons rejoined with the accuracy and precision of a single nucleotide. The splicing of pre-mRNAs is conducted by complexes of proteins and RNA molecules called spliceosomes.

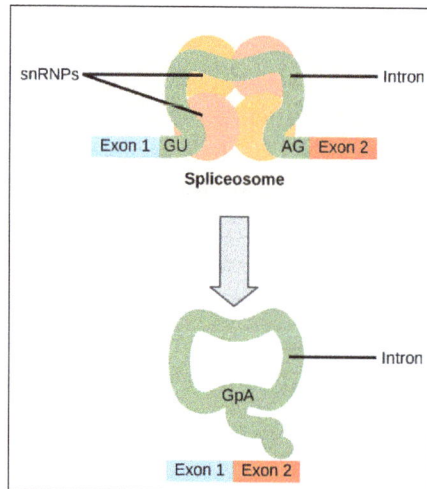

Pre-mRNA splicing.

Pre-mRNA splicing involves the precise removal of introns from the primary RNA transcript. The splicing process is catalyzed by large complexes called spliceosomes. Each spliceosome is composed of five subunits called snRNPs. The spliceseome's actions result in the splicing together of the two exons and the release of the intron in a lariat form.

Each spliceosome is composed of five subunits called snRNPs Each snRNP is itself a complex of proteins and a special type of RNA found only in the nucleus called snRNAs (small nuclear RNAs). Spliceosomes recognize sequences at the 5' end of the intron because introns always start with the nucleotides GU and they recognize sequences at the 3' end of the intron because they always end with the nucleotides AG. The spliceosome cleaves the pre-mRNA's sugar phosphate backbone at the G that starts the intron and then covalently attaches that G to an internal A nucleotide within the intron. Then the spliceosme connects the 3' end of the first exon to the 5' end of the following exon, cleaving the 3' end of the intron in the process. This results in the splicing together of the two exons and the release of the intron in a lariat form.

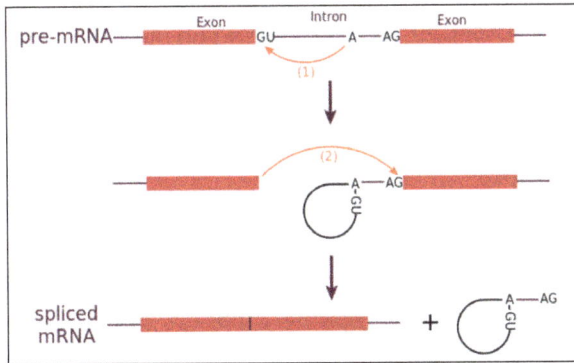

Mechanism of pre-mRNA splicing.

The snRNPs of the spliceosome were left out of this figure, but it shows the sites within the intron whose interactions are catalyzed by the spliceosome. Initially, the conserved G which starts an intron is cleaved from the 3' end of the exon upstream to it and the G is covalently attached to an internal A within the intron. Then the 3' end of the just-released exon is joined to the 5' end of the next exon, cleaving the bond that attaches the 3' end of the intron to its adjacent exon. This both joins the two exons and removes the intron in lariat form.

Processing of tRNAs and rRNAs

rRNA and tRNA are structural molecules that aid in protein synthesis but are not themselves translated into protein.

The tRNAs and rRNAs are structural molecules that have roles in protein synthesis; however, these RNAs are not themselves translated. In eukaryotes, pre-rRNAs are transcribed, processed, and assembled into ribosomes in the nucleolus, while pre-tRNAs are transcribed and processed in the nucleus and then released into the cytoplasm where they are linked to free amino acids for protein synthesis.

Ribosomal RNA (rRNA)

The four rRNAs in eukaryotes are first transcribed as two long precursor molecules. One contains just the pre-rRNA that will be processed into the 5S rRNA; the other spans the 28S, 5.8S, and 18S rRNAs. Enzymes then cleave the precursors into subunits corresponding to each rRNA. In bacteria, there are only three rRNAs and all are transcribed in one long precursor molecule that is cleaved into the individual rRNAs. Some of the bases of pre-rRNAs are methylated for added stability. Mature rRNAs make up 50-60% of each ribosome. Some of a ribosome's RNA molecules are purely structural, whereas others have catalytic or binding activities.

The eukaryotic ribosome is composed of two subunits: a large subunit (60S) and a small subunit (40S). The 60S subunit is composed of the 28S rRNA, 5.8S rRNA, 5S rRNA, and 50 proteins. The 40S subunit is composed of the 18S rRNA and 33 proteins. The bacterial ribosome is composed of two similar subunits, with slightly different components. The bacterial large subunit is called the 50S subunit and is composed of the 23S rRNA, 5S rRNA, and 31 proteins, while the bacterial small subunit is called the 30S subunit and is composed of the 16S rRNA and 21 proteins.

The two subunits join to constitute a functioning ribosome that is capable of creating proteins.

Transfer RNA (tRNA)

Each different tRNA binds to a specific amino acid and transfers it to the ribosome. Mature tRNAs take on a three-dimensional structure through intramolecular base-pairing to position the amino acid binding site at one end and the anticodon in an unbasepaired loop of nucleotides at the other end. The anticodon is a three-nucleotide sequence, unique to each different tRNA, that interacts with a messenger RNA (mRNA) codon through complementary base pairing.

There are different tRNAs for the 21 different amino acids. Most amino acids can be carried by more than one tRNA.

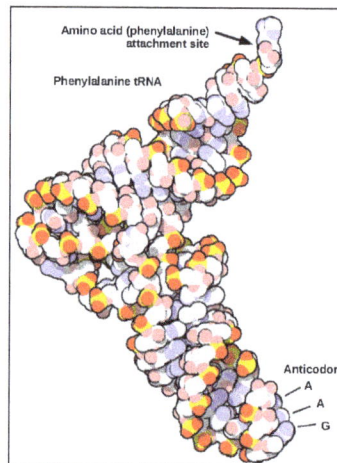

Structure of tRNA: This is a space-filling model of a tRNA molecule that adds the amino acid phenylalanine to a growing polypeptide chain. The anticodon AAG binds the codon UUC on the mRNA. The amino acid phenylalanine is attached to the other end of the tRNA.

In all organisms, tRNAs are transcribed in a pre-tRNA form that requires multiple processing steps before the mature tRNA is ready for use in translation. In bacteria, multiple tRNAs are often transcribed as a single RNA. The first step in their processing is the digestion of the RNA to release individual pre-tRNAs. In archaea and eukaryotes, each pre-tRNA is transcribed as a separate transcript.

The processing to convert the pre-tRNA to a mature tRNA involves five steps.

1. The 5′ end of the pre-tRNA, called the 5′ leader sequence, is cleaved off.

2. The 3′ end of the pre-tRNA is cleaved off.

3. In all eukaryote pre-tRNAs, but in only some bacterial and archaeal pre-tRNAs, a CCA sequence of nucleotides is added to the 3′ end of the pre-tRNA after the original 3′ end is trimmed off. Some bacteria and archaea pre-tRNAs already have the CCA encoded in their transcript immediately upstream of the 3′ cleavage site, so they don't need to add one. The CCA at the 3′ end of the mature tRNA will be the site at which the tRNA's amino acid will be added.

4. Multiple nucleotides in the pre-tRNA are chemically modified, altering their nitorgen bases. On average about 12 nucleotides are modified per tRNA. The most common modifications are the conversion of adenine (A) to pseudouridine (ψ), the conversion of adenine to inosine (I), and the conversion of uridine to dihydrouridine (D). But over 100 other modifications can occur.

5. A significant number of eukaryotic and archaeal pre-tRNAs have introns that have to be spliced out. Introns are rarer in bacterial pre-tRNAs, but do occur occasionally and are spliced out.

After processing, the mature pre-tRNA is ready to have its cognate amino acid attached. The cognate amino acid for a tRNA is the one specified by its anticodon. Attaching this amino acid is called charging the tRNA. In eukaryotes, the mature tRNA is generated in the nucleus, and then exported to the cytoplasm for charging.

Gene Expression

Gene expression is the process by which the genetic code - the nucleotide sequence - of a gene is used to direct protein synthesis and produce the structures of the cell. Genes that code for amino acid sequences are known as 'structural genes'.

The process of gene expression involves two main stages:

- Transcription: The production of messenger RNA (mRNA) by the enzyme RNA polymerase, and the processing of the resulting mRNA molecule.

- Translation: the use of mRNA to direct protein synthesis, and the subsequent post-translational processing of the protein molecule.

Some genes are responsible for the production of other forms of RNA that play a role in translation, including transfer RNA (tRNA) and ribosomal RNA (rRNA).

A structural gene involves a number of different components:

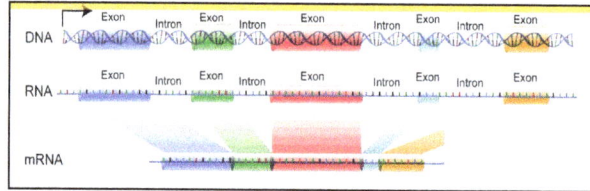

- Exons. Exons code for amino acids and collectively determine the amino acid sequence of the protein product. It is these portions of the gene that are represented in final mature mRNA molecule.

- Introns. Introns are portions of the gene that do not code for amino acids, and are removed (spliced) from the mRNA molecule before translation.

Gene Control Regions

- Start site. A start site for transcription.

- A promoter. A region a few hundred nucleotides 'upstream' of the gene (toward the 5' end). It is not transcribed into mRNA, but plays a role in controlling the transcription of the gene. Transcription factors bind to specific nucleotide sequences in the promoter region and assist in the binding of RNA polymerases.

- Enhancers. Some transcription factors (called activators) bind to regions called 'enhancers' that increase the rate of transcription. These sites may be thousands of nucleotides from the coding sequences or within an intron. Some enhancers are conditional and only work in the presence of other factors as well as transcription factors.

- Silencers. Some transcription factors (called repressors) bind to regions called 'silencers' that depress the rate of transcription.

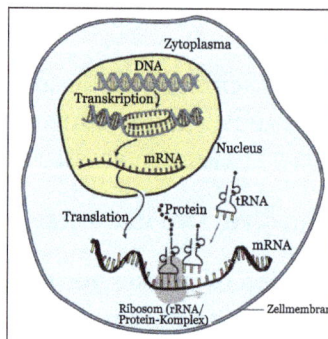

Transcription

Transcription is the process of RNA synthesis, controlled by the interaction of promoters and enhancers. Several different types of RNA are produced, including messenger RNA (mRNA), which specifies the sequence of amino acids in the protein product, plus transfer RNA (tRNA) and ribosomal RNA (rRNA), which play a role in the translation process.

Transcription involves four steps:

1. Initiation. The DNA molecule unwinds and separates to form a small open complex. RNA polymerase binds to the promoter of the template strand.

2. Elongation. RNA polymerase moves along the template strand, synthesising an mRNA molecule. In prokaryotes RNA polymerase is a holoenzyme consisting of a number of subunits, including a sigma factor (transcription factor) that recognises the promoter. In eukaryotes there are three RNA polymerases: I, II and III. The process includes a proofreading mechanism.

3. Termination. In prokaryotes there are two ways in which transcription is terminated. In Rho-dependent termination, a protein factor called "Rho" is responsible for disrupting the complex involving the template strand, RNA polymerase and RNA molecule. In Rho-independent termination, a loop forms at the end of the RNA molecule, causing it to detach itself. Termination in eukaryotes is more complicated, involving the addition of additional adenine nucleotides at the 3' of the RNA transcript (a process referred to as polyadenylation).

4. Processing. After transcription the RNA molecule is processed in a number of ways: introns are removed and the exons are spliced together to form a mature mRNA molecule consisting of a single protein-coding sequence. RNA synthesis involves the normal base pairing rules, but the base thymine is replaced with the base uracil.

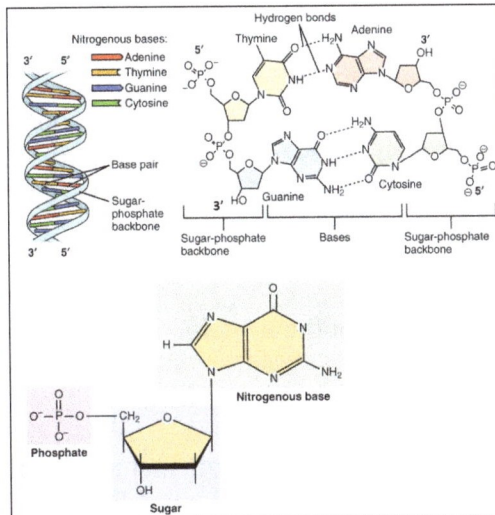

Translation

In translation the mature mRNA molecule is used as a template to assemble a series of amino acids to produce a polypeptide with a specific amino acid sequence. The complex in the cytoplasm at which this occurs is called a ribosome. Ribosomes are a mixture of ribosomal proteins and ribosomal RNA (rRNA), and consist of a large subunit and a small subunit.

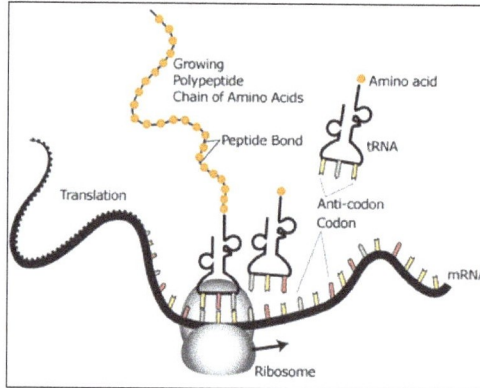

Translation involves four steps:

1. Initiation: The small subunit of the ribosome binds at the 5' end of the mRNA molecule and moves in a 3' direction until it meets a start codon (AUG). It then forms a complex with the large unit of the ribosome complex and an initiation tRNA molecule.

2. Elongation: Subsequent codons on the mRNA molecule determine which tRNA molecule linked to an amino acid binds to the mRNA. An enzyme peptidyl transferase links the amino acids together using peptide bonds. The process continues, producing a chain of amino acids as the ribosome moves along the mRNA molecule.

3. Termination: Translation in terminated when the ribosomal complex reached one or more stop codons (UAA, UAG, UGA). The ribosomal complex in eukaryotes is larger and more complicated than in prokaryotes. In addition, the processes of transcription and translation are divided in eukaryotes between the nucleus (transcription) and the cytoplasm (translation), which provides more opportunities for the regulation of gene expression.

4. Post-translation processing of the protein.

Regulation of Gene Expression

Regulation of gene expression, or gene regulation, includes a wide range of mechanisms that are used by cells to increase or decrease the production of specific gene products (protein or RNA). Sophisticated programs of gene expression are widely observed in biology, for example to trigger developmental pathways, respond to environmental stimuli, or adapt to new food sources. Virtually any step of gene expression can be modulated, from transcriptional initiation, to RNA processing, and to the post-translational modification of a protein. Often, one gene regulator controls another, and so on, in a gene regulatory network.

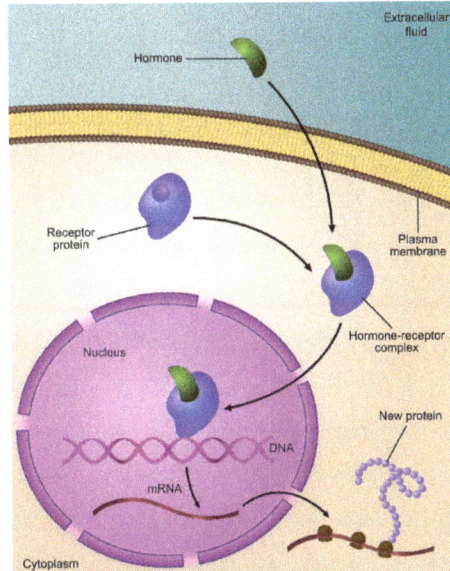

Regulation of gene expression by a hormone receptor.

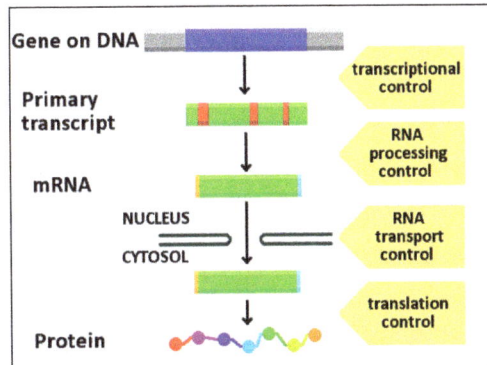

Stages in the DNA-mRNA-protein pathway expression can be controlled.

Gene regulation is essential for viruses, prokaryotes and eukaryotes as it increases the versatility and adaptability of an organism by allowing the cell to express protein when needed. Although as early as 1951, Barbara McClintock showed interaction between two genetic loci, Activator (*Ac*) and Dissociator (*Ds*), in the color formation of maize seeds, the first discovery of a gene regulation system is widely considered to be the identification in 1961 of the *lac* operon, discovered by François Jacob and Jacques Monod, in which some enzymes involved in lactose metabolism are expressed by *E. coli* only in the presence of lactose and absence of glucose.

In multicellular organisms, gene regulation drives cellular differentiation and morpho-genesis in the embryo, leading to the creation of different cell types that possess differ-ent gene expression profiles from the same genome sequence. Although this does not explain how gene regulation originated, evolutionary biologists include it as a partial explanation of how evolution works at a molecular level, and it is central to the science of evolutionary developmental biology ("evo-devo").

Regulated Stages of Gene Expression

Any step of gene expression may be modulated, from the DNA-RNA transcription step to post-translational modification of a protein. The following is a list of stages where gene expression is regulated, the most extensively utilised point is Transcription Initiation:

- Chromatin domains.

- Transcription.

- Post-transcriptional modification.

- RNA transport.

- Translation.

- mRNA degradation.

Modification of DNA

In eukaryotes, the accessibility of large regions of DNA can depend on its chromatin structure, which can be altered as a result of histone modifications directed by DNA methylation, ncRNA, or DNA-binding protein. Hence these modifications may up or down regulate the expression of a gene. Some of these modifications that regulate gene expression are inheritable and are referred to as epigenetic regulation.

Structural

Transcription of DNA is dictated by its structure. In general, the density of its packing is indicative of the frequency of transcription. Octameric protein complexes called nu-cleosomes are responsible for the amount of supercoiling of DNA, and these complex-es can be temporarily modified by processes such as phosphorylation or more perma-nently modified by processes such as methylation. Such modifications are considered to be responsible for more or less permanent changes in gene expression levels.

Chemical

Methylation of DNA is a common method of gene silencing. DNA is typically methylated by methyltransferase enzymes on cytosine nucleotides in a CpG dinucleotide sequence (also called "CpG islands" when densely clustered). Analysis of the pattern of methylation in a given region of DNA (which can be a promoter) can be achieved through a method called bisulfite mapping. Methylated cytosine residues are unchanged by the treatment, whereas unmethylated ones are changed to uracil. The differences are analyzed by DNA sequencing or by methods developed to quantify SNPs, such as Pyrosequencing (Biotage) or MassArray (Sequenom), measuring the relative amounts of C/T at the CG dinucleotide. Abnormal methylation patterns are thought to be involved in oncogenesis.

Histone acetylation is also an important process in transcription. Histone acetyltransferase enzymes (HATs) such as CREB-binding protein also dissociate the DNA from the histone complex, allowing transcription to proceed. Often, DNA methylation and histone deacetylation work together in gene silencing. The combination of the two seems to be a signal for DNA to be packed more densely, lowering gene expression.

Regulation of Transcription

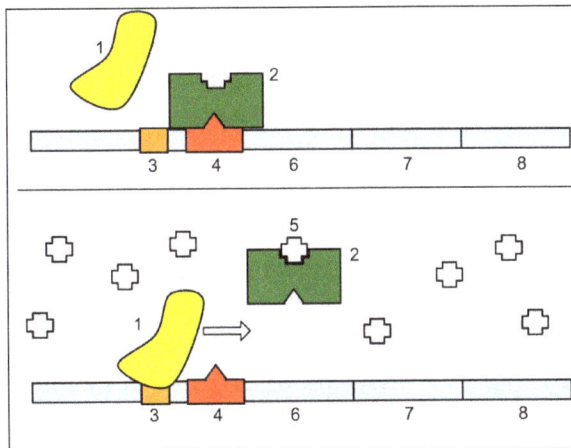

1: RNA Polymerase, 2: Repressor, 3: Promoter, 4: Operator, 5: Lactose, 6: lacZ, 7: lacY, 8: lacA.

Top: The gene is essentially turned off. There is no lactose to inhibit the repressor, so the repressor binds to the operator, which obstructs the RNA polymerase from binding to the promoter and making lactase. Bottom: The gene is turned on. Lactose is inhibiting the repressor, allowing the RNA polymerase to bind with the promoter, and express the genes, which synthesize lactase. Eventually, the lactase will digest all of the lactose, until there is none to bind to the repressor. The repressor will then bind to the operator, stopping the manufacture of lactase.

Regulation of transcription thus controls when transcription occurs and how much RNA is created. Transcription of a gene by RNA polymerase can be regulated by several mechanisms. Specificity factors alter the specificity of RNA polymerase for a given promoter or set of promoters, making it more or less likely to bind to them (i.e., sigma factors used in prokaryotic transcription). Repressors bind to the Operator, coding sequences on the DNA strand that are close to or overlapping the promoter region, impeding RNA polymerase's progress along the strand, thus impeding the expression of the gene.The image to the right demonstrates regulation by a repressor in the lac operon. General transcription factors position RNA polymerase at the start of a protein-coding sequence and then release the polymerase to transcribe the mRNA. Activators enhance the interaction between RNA polymerase and a particular promoter, encouraging the expression of the gene. Activators do this by increasing the attraction of RNA polymerase for the promoter, through interactions with subunits of the RNA polymerase or indirectly by changing the structure of the DNA. Enhancers are sites on the DNA helix that are bound by activators in order to loop the DNA bringing a specific promoter to the initiation complex. Enhancers are much more common in eukaryotes than prokaryotes, where only a few examples exist (to date). Silencers are regions of DNA sequences that, when bound by particular transcription factors, can silence expression of the gene.

Regulation of Transcription in Cancer

In vertebrates, the majority of gene promoters contain a CpG island with numerous CpG sites. When many of a gene's promoter CpG sites are methylated the gene becomes silenced. Colorectal cancers typically have 3 to 6 driver mutations and 33 to 66 hitchhiker or passenger mutations. However, transcriptional silencing may be of more importance than mutation in causing progression to cancer. For example, in colorectal cancers about 600 to 800 genes are transcriptionally silenced by CpG island methylation. Transcriptional repression in cancer can also occur by other epigenetic mechanisms, such as altered expression of microRNAs. In breast cancer, transcriptional repression of BRCA1 may occur more frequently by over-expressed microRNA-182 than by hypermethylation of the BRCA1 promoter.

Regulation of Transcription in Addiction

One of the cardinal features of addiction is its persistence. The persistent behavioral changes appear to be due to long-lasting changes, resulting from epigenetic alterations affecting gene expression, within particular regions of the brain. Drugs of abuse cause three types of epigenetic alteration in the brain. These are (1) histone acetylations and histone methylations, (2) DNA methylation at CpG sites, and (3) epigenetic downregulation or upregulation of microRNAs.

Chronic nicotine intake in mice alters brain cell epigenetic control of gene expression through acetylation of histones. This increases expression in the brain of the protein FosB, important in addiction. Cigarette addiction was also studied in about 16,000 humans, including never smokers, current smokers, and those who had quit smoking for up to 30 years. In blood cells, more than 18,000 CpG sites (of the roughly 450,000 analyzed CpG sites in the genome) had frequently altered methylation among current smokers. These CpG sites occurred in over 7,000 genes, or roughly a third of known human genes. The majority of the differentially methylated CpG sites returned to the level of never-smokers within five years of smoking cessation. However, 2,568 CpGs among 942 genes remained differentially methylated in former versus never smokers. Such remaining epigenetic changes can be viewed as "molecular scars" that may affect gene expression.

In rodent models, drugs of abuse, including cocaine, methampheamine, alcohol and tobacco smoke products, all cause DNA damage in the brain. During repair of DNA damages some individual repair events can alter the methylation of DNA and/or the acetylations or methylations of histones at the sites of damage, and thus can contribute to leaving an epigenetic scar on chromatin.

Such epigenetic scars likely contribute to the persistent epigenetic changes found in addiction.

Post-transcriptional Regulation

After the DNA is transcribed and mRNA is formed, there must be some sort of regulation on how much the mRNA is translated into proteins. Cells do this by modulating the capping, splicing, addition of a Poly(A) Tail, the sequence-specific nuclear export rates, and, in several contexts, sequestration of the RNA transcript. These processes occur in eukaryotes but not in prokaryotes. This modulation is a result of a protein or transcript that, in turn, is regulated and may have an affinity for certain sequences.

Three Prime Untranslated Regions and microRNAs

Three prime untranslated regions (3'-UTRs) of messenger RNAs (mRNAs) often contain regulatory sequences that post-transcriptionally influence gene expression. Such 3'-UTRs often contain both binding sites for microRNAs (miRNAs) as well as for regulatory proteins. By binding to specific sites within the 3'-UTR, miRNAs can decrease gene expression of various mRNAs by either inhibiting translation or directly causing degradation of the transcript. The 3'-UTR also may have silencer regions that bind repressor proteins that inhibit the expression of a mRNA.

The 3'-UTR often contains miRNA response elements (MREs). MREs are sequences to which miRNAs bind. These are prevalent motifs within 3'-UTRs. Among all regulatory motifs within the 3'-UTRs (e.g. including silencer regions), MREs make up about half of the motifs.

An archive of miRNA sequences and annotations, listed 28,645 entries in 233 biologic species. Of these, 1,881 miRNAs were in annotated human miRNA loci. miRNAs were predicted to have an average of about four hundred target mRNAs (affecting expression of several hundred genes). Freidman et al. estimate that >45,000 miRNA target sites within human mRNA 3'-UTRs are conserved above background levels, and >60% of human protein-coding genes have been under selective pressure to maintain pairing to miRNAs.

Direct experiments show that a single miRNA can reduce the stability of hundreds of unique mRNAs. Other experiments show that a single miRNA may repress the production of hundreds of proteins, but that this repression often is relatively mild (less than 2-fold).

The effects of miRNA dysregulation of gene expression seem to be important in cancer. For instance, in gastrointestinal cancers, identified nine miRNAs as epigenetically altered and effective in down-regulating DNA repair enzymes.

The effects of miRNA dysregulation of gene expression also seem to be important in neuropsychiatric disorders, such as schizophrenia, bipolar disorder, major depressive disorder, Parkinson's disease, Alzheimer's disease and autism spectrum disorders.

Regulation of Translation

The translation of mRNA can also be controlled by a number of mechanisms, mostly at the level of initiation. Recruitment of the small ribosomal subunit can indeed be modulated by mRNA secondary structure, antisense RNA binding, or protein binding. In both prokaryotes and eukaryotes, a large number of RNA binding proteins exist, which often are directed to their target sequence by the secondary structure of the transcript, which may change depending on certain conditions, such as temperature or presence of a ligand (aptamer). Some transcripts act as ribozymes and self-regulate their expression.

Examples of Gene Regulation

- Enzyme induction is a process in which a molecule (e.g., a drug) induces (i.e., initiates or enhances) the expression of an enzyme.

- The induction of heat shock proteins in the fruit fly *Drosophila melanogaster*.

- The Lac operon is an interesting example of how gene expression can be regulated.

- Viruses, despite having only a few genes, possess mechanisms to regulate their gene expression, typically into an early and late phase, using collinear systems regulated by anti-terminators (lambda phage) or splicing modulators (HIV).

- Gal4 is a transcriptional activator that controls the expression of GAL1, GAL7, and GAL10 (all of which code for the metabolic of galactose in yeast). The GAL4/UAS system has been used in a variety of organisms across various phyla to study gene expression.

Developmental biology

A large number of studied regulatory systems come from developmental biology. Examples include:

- The colinearity of the Hox gene cluster with their nested antero-posterior patterning.

- Pattern generation of the hand (digits - interdigits): The gradient of sonic hedgehog (secreted inducing factor) from the zone of polarizing activity in the limb, which creates a gradient of active Gli3, which activates Gremlin, which inhibits BMPs also secreted in the limb, results in the formation of an alternating pattern of activity as a result of this reaction-diffusion system.

- Somitogenesis is the creation of segments (somites) from a uniform tissue (Pre-somitic Mesoderm). They are formed sequentially from anterior to posterior. This is achieved in amniotes possibly by means of two opposing gradients, Retinoic acid in the anterior (wavefront) and Wnt and Fgf in the posterior, coupled to an oscillating pattern (segmentation clock) composed of FGF + Notch and Wnt in antiphase.

- Sex determination in the soma of a Drosophila requires the sensing of the ratio of autosomal genes to sex chromosome-encoded genes, which results in the production of sexless splicing factor in females, resulting in the female isoform of doublesex.

Circuitry

Up-regulation and Down-regulation

Up-regulation is a process that occurs within a cell triggered by a signal (originating internal or external to the cell), which results in increased expression of one or more genes and as a result the protein(s) encoded by those genes. Conversely, down-regulation is a process resulting in decreased gene and corresponding protein expression.

- Up-regulation occurs, for example, when a cell is deficient in some kind of receptor. In this case, more receptor protein is synthesized and transported to the membrane of the cell and, thus, the sensitivity of the cell is brought back to normal, reestablishing homeostasis.

- Down-regulation occurs, for example, when a cell is overstimulated by a neurotransmitter, hormone, or drug for a prolonged period of time, and the expression of the receptor protein is decreased in order to protect the cell.

Inducible vs. Repressible Systems

Gene Regulation can be summarized by the response of the respective system:

- Inducible systems - An inducible system is off unless there is the presence of some molecule (called an inducer) that allows for gene expression. The molecule is said to "induce expression". The manner by which this happens is dependent on the control mechanisms as well as differences between prokaryotic and eukaryotic cells.

- Repressible systems - A repressible system is on except in the presence of some molecule (called a corepressor) that suppresses gene expression. The molecule is said to "repress expression". The manner by which this happens is dependent on the control mechanisms as well as differences between prokaryotic and eukaryotic cells.

The GAL4/UAS system is an example of both an inducible and repressible system. Gal4 binds an upstream activation sequence (UAS) to activate the transcription of the GAL1/GAL7/GAL10 cassette. On the other hand, a MIG1 response to the presence of glucose can inhibit GAL4 and therefore stop the expression of the GAL1/GAL7/GAL10 cassette.

Theoretical Circuits

- Repressor/Inducer: an activation of a sensor results in the change of expression of a gene.

- Negative feedback: the gene product downregulates its own production directly or indirectly, which can result in:

  ○ Keeping transcript levels constant/proportional to a factor.

  ○ Inhibition of run-away reactions when coupled with a positive feedback loop.

  ○ Creating an oscillator by taking advantage in the time delay of transcription and translation, given that the mRNA and protein half-life is shorter.

- Positive feedback: the gene product upregulates its own production directly or indirectly, which can result in:

  ○ Signal amplification.

  ○ Bistable switches when two genes inhibit each other and both have positive feedback.

  ○ Pattern generation.

Study Methods

In general, most experiments investigating differential expression used whole cell extracts of RNA, called steady-state levels, to determine which genes changed and by how much. These are, however, not informative of where the regulation has occurred and may mask conflicting regulatory processes, but it is still the most commonly analysed (quantitative PCR and DNA microarray).

When studying gene expression, there are several methods to look at the various stages. In eukaryotes these include:

- The local chromatin environment of the region can be determined by ChIP-chip analysis by pulling down RNA Polymerase II, Histone 3 modifications, Tritho-rax-group protein, Polycomb-group protein, or any other DNA-binding element to which a good antibody is available.

- Epistatic interactions can be investigated by synthetic genetic array analysis.

- Due to post-transcriptional regulation, transcription rates and total RNA levels differ significantly. To measure the transcription rates nuclear run-on assays can be done and newer high-throughput methods are being developed, using thiol labelling instead of radioactivity.

- Only 5% of the RNA polymerised in the nucleus exits, and not only introns, abortive products, and non-sense transcripts are degradated. Therefore, the differences in nuclear and cytoplasmic levels can be see by separating the two fractions by gentle lysis.

- Alternative splicing can be analysed with a splicing array or with a tiling array.

- All in vivo RNA is complexed as RNPs. The quantity of transcripts bound to specific protein can be also analysed by RIP-Chip. For example, DCP2 will give an indication of sequestered protein; ribosome-bound gives and indication of transcripts active in transcription (although a more dated method, called polysome fractionation, is still popular in some labs).

- Protein levels can be analysed by Mass spectrometry, which can be compared only to quantitative PCR data, as microarray data is relative and not absolute.

- RNA and protein degradation rates are measured by means of transcription inhibitors (actinomycin D or α-amanitin) or translation inhibitors (Cycloheximide), respectively.

Measuring Gene Expression

Measuring gene expression is an important part of many life sciences, as the ability to quantify the level at which a particular gene is expressed within a cell, tissue or organism can provide a lot of valuable information. For example, measuring gene expression can:

- Identify viral infection of a cell (viral protein expression).

- Determine an individual's susceptibility to cancer (oncogene expression).

- Find if a bacterium is resistant to penicillin (beta-lactamase expression).

Similarly, the analysis of the location of protein expression is a powerful tool, and this can be done on an organismal or cellular scale. Investigation of localization is particularly important for the study of development in multicellular organisms and as an indicator of protein function in single cells. Ideally, measurement of expression is done by detecting the final gene product (for many genes, this is the protein); however, it is often easier to detect one of the precursors, typically mRNA and to infer gene-expression levels from these measurements.

mRNA Quantification

Levels of mRNA can be quantitatively measured by northern blotting, which provides size and sequence information about the mRNA molecules. A sample of RNA is separated on an agarose gel and hybridized to a radioactively labeled RNA probe that is complementary to the target sequence. The radiolabeled RNA is then detected by an autoradiograph. Because the use of radioactive reagents makes the procedure time consuming and potentially dangerous, alternative labeling and detection methods, such as digoxigenin and biotin chemistries, have been developed. Perceived disadvantages of Northern blotting are that large quantities of RNA are required and that quantification may not be completely accurate, as it involves measuring band strength in an image of a gel. On the other hand, the additional mRNA size information from the Northern blot allows the discrimination of alternately spliced transcripts.

Another approach for measuring mRNA abundance is RT-qPCR. In this technique, reverse transcription is followed by quantitative PCR. Reverse transcription first generates a DNA template from the mRNA; this single-stranded template is called cDNA. The cDNA template is then amplified in the quantitative step, during which the fluorescence emitted by labeled hybridization probes or intercalating dyes changes as the DNA amplification process progresses. With a carefully constructed standard curve, qPCR can produce an absolute measurement of the number of copies of original mRNA, typically in units of copies per nanolitre of homogenized tissue or copies per cell. qPCR is very sensitive (detection of a single mRNA molecule is theoretically possible), but can be expensive depending on the type of reporter used; fluorescently labeled oligonucleotide probes are more expensive than non-specific intercalating fluorescent dyes.

For expression profiling, or high-throughput analysis of many genes within a sample, quantitative PCR may be performed for hundreds of genes simultaneously in the case of low-density arrays. A second approach is the hybridization microarray.

A single array or "chip" may contain probes to determine transcript levels for every known gene in the genome of one or more organisms. Alternatively, "tag based" technologies like Serial analysis of gene expression (SAGE) and RNA-Seq, which can provide a relative measure of the cellular concentration of different mRNAs, can be used. An advantage of tag-based methods is the "open architecture", allowing for the exact measurement of any transcript, with a known or unknown sequence. Next-generation sequencing (NGS) such as RNA-Seq is another approach, producing vast quantities of sequence data that can be matched to a reference genome. Although NGS is comparatively time-consuming, expensive, and resource-intensive, it can identify single-nucleotide polymorphisms, splice-variants, and novel genes, and can also be used to profile expression in organisms for which little or no sequence information is available.

Protein Quantification

For genes encoding proteins, the expression level can be directly assessed by a number of methods with some clear analogies to the techniques for mRNA quantification.

The most commonly used method is to perform a Western blot against the protein of interest—this gives information on the size of the protein in addition to its identity. A sample (often cellular lysate) is separated on a polyacrylamide gel, transferred to a membrane and then probed with an antibody to the protein of interest. The antibody can either be conjugated to a fluorophore or to horseradish peroxidase for imaging and/or quantification. The gel-based nature of this assay makes quantification less accurate, but it has the advantage of being able to identify later modifications to the protein, for example proteolysis or ubiquitination, from changes in size.

mRNA-protein Correlation

Quantification of protein and mRNA permits a correlation of the two levels. The question of how well protein levels correlate with their corresponding transcript levels is highly debated and depends on multiple factors. Regulation on each step of gene expression can impact the correlation, as shown for regulation of translation or protein stability. Post-translational factors, such as protein transport in highly polar cells, can influence the measured mRNA-protein correlation as well.

Localisation

Analysis of expression is not limited to quantification; localisation can also be determined. mRNA can be detected with a suitably labelled complementary mRNA strand and protein can be detected via labelled antibodies. The probed sample is then observed by microscopy to identify where the mRNA or protein is.

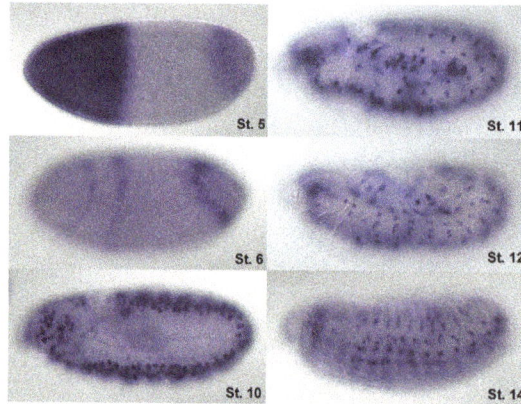

In situ-hybridization of Drosophila embryos at different developmental stages for the mRNA
responsible for the expression of hunchback. High intensity of blue color marks places
with high hunchback mRNA quantity.

By replacing the gene with a new version fused to a green fluorescent protein (or similar) marker, expression may be directly quantified in live cells. This is done by imaging
using a fluorescence microscope. It is very difficult to clone a GFP-fused protein into its
native location in the genome without affecting expression levels so this method often
cannot be used to measure endogenous gene expression. It is, however, widely used to
measure the expression of a gene artificially introduced into the cell, for example via an
expression vector. It is important to note that by fusing a target protein to a fluorescent
reporter the protein's behavior, including its cellular localization and expression level,
can be significantly changed.

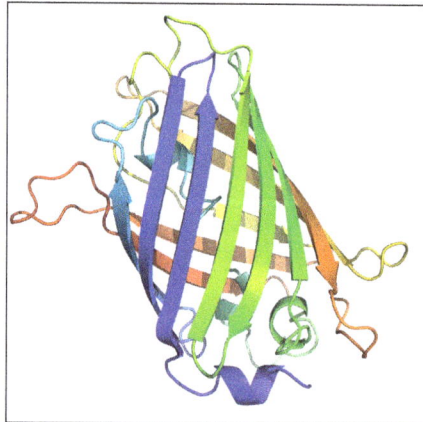

The three-dimensional structure of green fluorescent protein. The residues
in the centre of the "barrel" are responsible for production of green light
after exposing to higher energetic blue light.

The enzyme-linked immunosorbent assay works by using antibodies immobilised on a
microtiter plate to capture proteins of interest from samples added to the well. Using
a detection antibody conjugated to an enzyme or fluorophore the quantity of bound
protein can be accurately measured by fluorometric or colourimetric detection. The

detection process is very similar to that of a Western blot, but by avoiding the gel steps more accurate quantification can be achieved.

Expression System

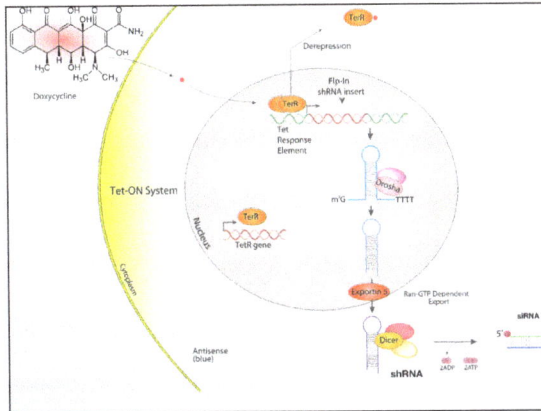

Tet-ON inducible shRNA system.

An expression system is a system specifically designed for the production of a gene product of choice. This is normally a protein although may also be RNA, such as tRNA or a ribozyme. An expression system consists of a gene, normally encoded by DNA, and the molecular machinery required to transcribe the DNA into mRNA and translate the mRNA into protein using the reagents provided. In the broadest sense this includes every living cell but the term is more normally used to refer to expression as a laboratory tool. An expression system is therefore often artificial in some manner. Expression systems are, however, a fundamentally natural process. Viruses are an excellent example where they replicate by using the host cell as an expression system for the viral proteins and genome.

Inducible Expression

Doxycycline is also used in "Tet-on" and "Tet-off" tetracycline controlled transcriptional activation to regulate transgene expression in organisms and cell cultures.

In Nature

In addition to these biological tools, certain naturally observed configurations of DNA (genes, promoters, enhancers, repressors) and the associated machinery itself are referred to as an expression system. This term is normally used in the case where a gene or set of genes is switched on under well defined conditions, for example, the simple repressor switch expression system in Lambda phage and the lac operator system in bacteria. Several natural expression systems are directly used or modified and used for artificial expression systems such as the Tet-on and Tet-off expression system.

Gene Expression Profiling

Gene expression profiling measures which genes are being expressed in a cell at any given moment. This method can measure thousands of genes at a time; some experiments can measure the entire genome at once. Gene expression profiling measures mRNA levels, showing the pattern of genes expressed by a cell at the transcription level. This often means measuring relative mRNA amounts in two or more experimental conditions, then assessing which conditions resulted in specific genes being expressed.

Gene expression profiling is used by a variety of biomedical researchers, from molecular biologists to environmental toxicologists. This technology can provide accurate information on gene expression, towards countless experimental goals.

Different techniques are used to determine gene expression. These include DNA microarrays and sequencing technologies. The former measures the activity of specific genes of interest and the latter enables researchers to determine all active genes in a cell.

Once a genome has been sequenced, we know what potential a cell has—what characteristics and function it might have—based on the genes it contains. However, sequencing the genome does not tell us which genes a cell is expressing, or the functions or processes it is carrying out at any given moment. To determine these, we need to work out its gene expression profile. If a gene is being used to make mRNA, it is considered 'on'; if it is not being used to make mRNA, it is considered 'off'.

A gene expression profile tells us how a cell is functioning at a specific time. This is because cell gene expression is influenced by external and internal stimuli, including whether the cell is dividing, what factors are present in the cell's environment, the signals it is receiving from other cells, and even the time of day.

Uses of Gene Expression Profiling

Gene expression profiling enables you to investigate the effects of different conditions on gene expression by altering the environment to which the cell is exposed, and determining which genes are expressed. Alternatively, if you already know a gene is involved in a certain cell behavior, gene expression profiling helps you to determine whether a cell is carrying out this function. For example, certain genes are known to be involved in cell division; if these genes are active in a cell, you can tell the cell is undergoing division, or whether a cell is differentiated.

Gene expression profiling is often used in hypothesis generation. If very little is known about when and why a gene will be expressed, expression profiling under different conditions can help design a hypothesis to test in future experiments. For example, if gene A is expressed only when the cell is exposed to other cells, this gene may be involved in intercellular communication.

Gene profiling can also investigate the effect of drug-like molecules on cellular response. You could identify the gene markers of drug metabolism, or determine whether cells express genes known to be involved in response to toxic environments when exposed to the drug.

Gene profiling can also be used as a diagnostic tool. If cancerous cells express higher levels of certain genes, and these genes code for a protein receptor, this receptor may be involved in the cancer, and targeting it with a drug might treat the disease. Gene expression profiling might then be a key diagnostic tool for people with this cancer.

Different Types of Gene Expression Profiling

RNA expression patterns are key to predicting and classifying human disease based on specific biomarkers. To understand cellular responses, we must determine how gene expression changes are affected in relation to external stimuli, different environmental conditions and genetic lesions.

Transcriptome sequencing, using next-generation sequencing, lets us discover differentially expressed genes without requiring knowledge of which genes are involved.

Protein-coding RNAs are an important source of information, though non-coding RNA is also significant. Next-generation RNA sequencing enables such analysis, along with:

- 'Digital counting' of RNA molecules for highly quantitative and precise measurements.

- Dynamic ranges to fully capture relevant biological changes.

- The discovery of unknown RNAs (novel transcripts, splice variants, and gene fusions).

- The capture of all RNA types (poly-A+, long non-coding RNA and gene fusions) in a single assay.

- Opportunities to focus, going from complete transcriptome analysis to a handful of pre-selected RNA sequences to balance experimental cost and ease of analysis with discovery potential.

Quantification of mRNA using qPCR can be done using Applied Biosystems™ TaqMan® probe–based analysis and Applied Biosystems™ SYBR™ Green dye–based analysis plus using digital PCR, as discussed subsequently.

qPCR is the gold-standard technique for validating differential gene expression profiles, and enables:

- Quantitation of gene products.

- Microarray validation.

- Pathway analysis.

- Studies of developmental biology.

- Quality control and assay validation.

- siRNA/RNAi experiments.

- Low-fold copy number discrimination (down to two-fold).

- Mid- to high-throughput profiling using the OpenArray platform.

Although qPCR is useful for detecting gene expression changes of two-fold or more, a different approach is needed for measuring less than two-fold changes. Digital PCR (dPCR) can be used to resolve low-fold gene expression changes. dPCR enables:

- Absolute quantification of nucleic acid standards and next-generation sequencing libraries.

- Rare target detection.

- Enrichment and separation of mixtures.

Protein Structure

A protein's primary structure is the unique sequence of amino acids in each polypeptide chain that makes up the protein. not a structure. But, because the final protein structure ultimately depends on this sequence, this was called the primary structure of the polypeptide chain. For example, the pancreatic hormone insulin has two polypeptide chains, A and B.

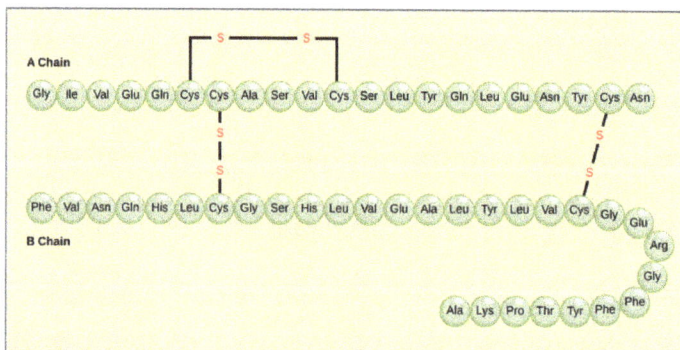

Primary structure: The A chain of insulin is 21 amino acids long and the
B chain is 30 amino acids long, and each sequence is unique to the insulin protein.

The gene, or sequence of DNA, ultimately determines the unique sequence of amino acids in each peptide chain. A change in nucleotide sequence of the gene's coding region may lead to a different amino acid being added to the growing polypeptide chain, causing a change in protein structure and therefore function.

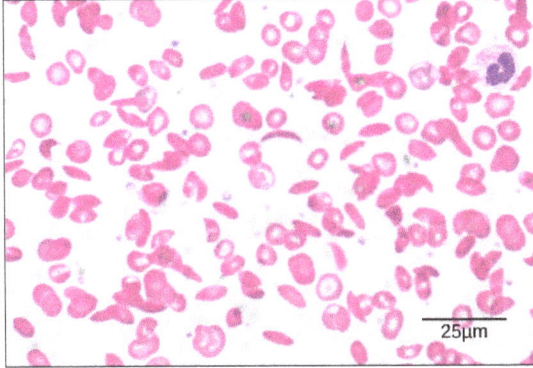

Sickle cell disease: Sickle cells are crescent shaped, while normal cells are disc-shaped.

The oxygen-transport protein hemoglobin consists of four polypeptide chains, two identical α chains and two identical β chains. In sickle cell anemia, a single amino substitution in the hemoglobin β chain causes a change the structure of the entire protein. When the amino acid glutamic acid is replaced by valine in the β chain, the polypeptide folds into an slightly-different shape that creates a dysfunctional hemoglobin protein. So, just one amino acid substitution can cause dramatic changes. These dysfunctional hemoglobin proteins, under low-oxygen conditions, start associating with one another, forming long fibers made from millions of aggregated hemoglobins that distort the red blood cells into crescent or "sickle" shapes, which clog arteries. People affected by the disease often experience breathlessness, dizziness, headaches, and abdominal pain.

Secondary Structure

A protein's secondary structure is whatever regular structures arise from interactions between neighboring or near-by amino acids as the polypeptide starts to fold into its functional three-dimensional form. Secondary structures arise as H bonds form between local groups of amino acids in a region of the polypeptide chain. Rarely does a single secondary structure extend throughout the polypeptide chain. It is usually just in a section of the chain. The most common forms of secondary structure are the α-helix and β-pleated sheet structures and they play an important structural role in most globular and fibrous proteins.

In the α-helix chain, the hydrogen bond forms between the oxygen atom in the polypeptide backbone carbonyl group in one amino acid and the hydrogen atom in the polypeptide backbone amino group of another amino acid that is four amino acids farther along the chain. This holds the stretch of amino acids in a right-handed coil. Every helical turn in an alpha helix has 3.6 amino acid residues. The R groups (the side chains) of the polypeptide protrude out from the α-helix chain and are not involved in the H bonds that maintain the α-helix structure.

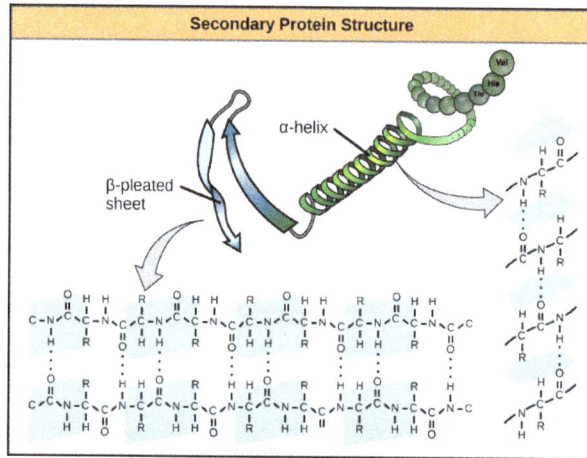

Secondary structure: The α-helix and β-pleated sheet form because of hydrogen bonding between carbonyl and amino groups in the peptide backbone. Certain amino acids have a propensity to form an α-helix, while others have a propensity to form a β-pleated sheet.

In β-pleated sheets, stretches of amino acids are held in an almost fully-extended conformation that "pleats" or zig-zags due to the non-linear nature of single C-C and C-N covalent bonds. β-pleated sheets never occur alone. They have to held in place by other β-pleated sheets. The stretches of amino acids in β-pleated sheets are held in their pleated sheet structure because hydrogen bonds form between the oxygen atom in a polypeptide backbone carbonyl group of one β-pleated sheet and the hydrogen atom in a polypeptide backbone amino group of another β-pleated sheet. The β-pleated sheets which hold each other together align parallel or antiparallel to each other. The R groups of the amino acids in a β-pleated sheet point out perpendicular to the hydrogen bonds holding the β-pleated sheets together, and are not involved in maintaining the β-pleated sheet structure.

Tertiary Structure

The tertiary structure of a polypeptide chain is its overall three-dimensional shape, once all the secondary structure elements have folded together among each other. Interactions between polar, nonpolar, acidic, and basic R group within the polypeptide chain create the complex three-dimensional tertiary structure of a protein. When protein folding takes place in the aqueous environment of the body, the hydrophobic R groups of nonpolar amino acids mostly lie in the interior of the protein, while the hydrophilic R groups lie mostly on the outside. Cysteine side chains form disulfide linkages in the presence of oxygen, the only covalent bond forming during protein folding. All of these interactions, weak and strong, determine the final three-dimensional shape of the protein. When a protein loses its three-dimensional shape, it will no longer be functional.

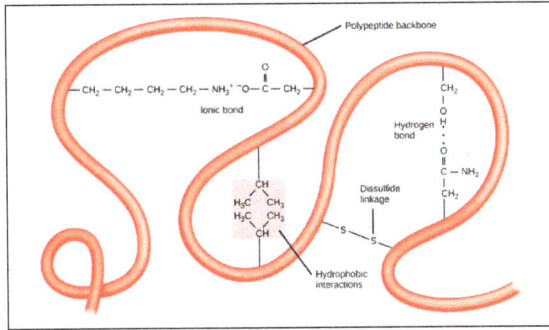

Tertiary structure: The tertiary structure of proteins is determined by hydrophobic interactions, ionic bonding, hydrogen bonding, and disulfide linkages.

Quaternary Structure

The quaternary structure of a protein is how its subunits are oriented and arranged with respect to one another. As a result, quaternary structure only applies to multi-subunit proteins; that is, proteins made from one than one polypeptide chain. Proteins made from a single polypeptide will not have a quaternary structure.

In proteins with more than one subunit, weak interactions between the subunits help to stabilize the overall structure. Enzymes often play key roles in bonding subunits to form the final, functioning protein.

For example, insulin is a ball-shaped, globular protein that contains both hydrogen bonds and disulfide bonds that hold its two polypeptide chains together. Silk is a fibrous protein that results from hydrogen bonding between different β-pleated chains.

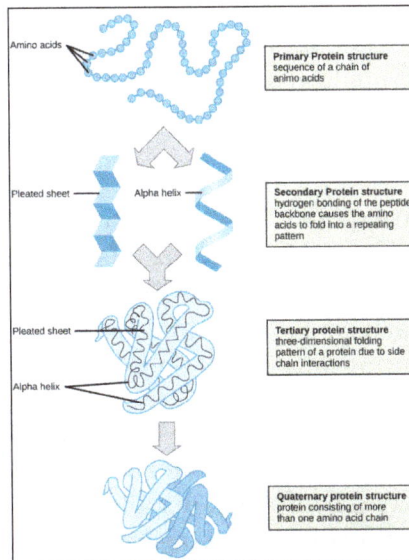

Four levels of protein structure: The four levels of protein structure can be observed in these illustrations.

Denaturation and Protein Folding

Each protein has its own unique shape. If the temperature or pH of a protein's environment is changed, or if it is exposed to chemicals, these interactions may be disrupted, causing the protein to lose its three-dimensional structure and turn back into an unstructured string of amino acids. When a protein loses its higher-order structure, but not its primary sequence, it is said to be denatured. Denatured proteins are usually non-functional.

For some proteins, denaturation can be reversed. Since the primary structure of the polypeptide is still intact (the amino acids haven't split up), it may be able to re-fold into its functional form if it's returned to its normal environment. Other times, however, denaturation is permanent. One example of irreversible protein denaturation is when an egg is fried. The albumin protein in the liquid egg white becomes opaque and solid as it is denatured by the heat of the stove, and will not return to its original, raw-egg state even when cooled down.

Since these proteins can go from unstructured to folded all by themselves, their amino acid sequences must contain all the information needed for folding. Many proteins don't fold by themselves, but instead get assistance from chaperone proteins (chaperonins).

Protein Stability

Due to the nature of the weak interactions controlling the three-dimensional structure, proteins are very sensitive molecules. The term native state is used to describe the protein in its most stable natural conformation in situ. This native state can be disrupted by a number of external stress factors including temperature, pH, removal of water, presence of hydrophobic surfaces, presence of metal ions and high shear. The loss of secondary, tertiary or quaternary structure due to exposure to a stress factor is called denaturation. Denaturation results in unfolding of the protein into a random or misfolded shape.

A denatured protein can have quite a different activity profile than the protein in its native form, usually losing biological function. In addition to becoming denatured, proteins can also form aggregates under certain stress conditions. Aggregates are often produced during the manufacturing process and are typically undesirable, largely due to the possibility of them causing adverse immune responses when administered.

In addition to these physical forms of protein degradation, it is also important to be aware of the possible pathways of protein chemical degradation. These include oxidation, deamidation, peptide-bond hydrolysis, disulfide-bond reshuffling and cross-linking. The methods used in the processing and the formulation of proteins, including any lyophilization step, must be carefully examined to prevent degradation and to increase the stability of the protein biopharmaceutical both in storage and during drug delivery.

Protein Structure Analysis

The complexities of protein structure make the elucidation of a complete protein structure extremely difficult even with the most advanced analytical equipment. An amino acid analyzer can be used to determine which amino acids are present and the molar ratios of each. The sequence of the protein can then be analyzed by means of peptide mapping and the use of Edman degradation or mass spectroscopy. This process is routine for peptides and small proteins, but becomes more complex for large multimeric proteins.

Peptide mapping generally entails treatment of the protein with different protease enzymes in order to chop up the sequence into smaller peptides at specific cleavage sites. Two commonly used enzymes are trypsin and chymotrypsin. Mass spectroscopy has become an invaluable tool for the analysis of enzyme digested proteins, by means of peptide fingerprinting methods and database searching. Edman degradation involves the cleavage, separation and identification of one amino acid at a time from a short peptide, starting from the N-terminus.

One method used to characterize the secondary structure of a protein is circular dichroism spectroscopy (CD). The different types of secondary structure, α-helix, ß-sheet and random coil, all have characteristic circular dichroism spectra in the far-uv region of the spectrum (190-250 nm). These spectra can be used to approximate the fraction of the entire protein made up of each type of structure.

A more complete, high-resolution analysis of the three-dimensional structure of a protein is carried out using X-ray crystallography or nuclear magnetic resonance (NMR) analysis. To determine the three-dimensional structure of a protein by X-ray diffraction, a large, well-ordered single crystal is required. X-ray diffraction allows measurement of the short distances between atoms and yields a three-dimensional electron density map, which can be used to build a model of the protein structure.

The use of NMR to determine the three-dimensional structure of a protein has some advantages over X-ray diffraction in that it can be carried out in solution and thus the protein is free of the constraints of the crystal lattice. The two-dimensional NMR techniques generally used are NOESY, which measures the distances between atoms through space, and COESY, which measures distances through bonds.

Protein Structure Stability Analysis

Many different techniques can be used to determine the stability of a protein. For the analysis of unfolding of a protein, spectroscopic methods such as fluorescence, UV, infrared and CD can be used. Thermodynamic methods such as differential scanning calorimetry (DSC) can be useful in determining the effect of temperature on protein stability. Comparative peptide-mapping (usually using LC/MS) is an extremely valuable tool in determining chemical changes in a protein such as oxidation or deamidation.

HPLC is also an invaluable means of analyzing the purity of a protein. Other analytical methods such as SDS-PAGE, iso-electric focusing and capillary electrophoresis can also be used to determine protein stability, and a suitable bioassay should be used to determine the potency of a protein biopharmaceutical. The state of aggregation can be determined by following "particle" size and arrayed instruments are now available to follow this over time under various conditions.

References

- Noncodingdna: nlm.nih.gov, Retrieved 13 July, 2019

- Dna-replication: lumenlearning.com, Retrieved 6 February, 2019

- Translation-dna-to-mrna-to-protein: nature.com, Retrieved 16 August, 2019

- RNA, science: britannica.com, Retrieved 7 March, 2019

- Ribosomal-rna: biologydictionary.net, Retrieved 18 May, 2019

- Lewin's genes X. Lewin, Benjamin., Krebs, Jocelyn E., Kilpatrick, Stephen T., Goldstein, Elliott S., Lewin, Benjamin. (10th ed.). Sudbury, Mass.: Jones and Bartlett. 2011. ISBN 9780763766320

- Trna: biologydictionary.net, Retrieved 22 March, 2019

- Transcription: biologydictionary.net, Retrieved 17 August, 2019

- Rna-processing-in-eukaryotes, biology: lumenlearning.com, Retrieved 26 May, 2019

- López-Lastra M, Rivas A, Barría MI (2005). "Protein synthesis in eukaryotes: the growing biological relevance of cap-independent translation initiation". Biological Research. 38 (2–3): 121–46. doi:10.4067/S0716-97602005000200003

- Protein-structure: lumenlearning.com, Retrieved 3 January, 2019

- Orders-of-protein-structure, proteins-and-amino-acids, macromolecules, biology, science: khanacademy.org, Retrieved 14 April, 2019

- Protein-structure: particlesciences.com, Retrieved 26 June, 2019

3

Human Genome and Chromosome

The genome of Homo sapiens, which is made up of 23 pairs of chromosomes is known as the human genome. It consists of both protein coding DNA genes and non-coding DNA genes. A chromosome is a DNA molecule that contains a part or all of the genetic material of an organism. The topics elaborated in this chapter will help in gaining a better perspective about the human genome, the epigenome and the chromosomes.

Human Genome

Human Genome is all of the approximately three billion base pairs of deoxyribonucleic acid (DNA) that make up the entire set of chromosomes of the human organism. The human genome includes the coding regions of DNA, which encode all the genes (between 20,000 and 25,000) of the human organism, as well as the noncoding regions of DNA, which do not encode any genes. By 2003 the DNA sequence of the entire human genome was known.

The human genome, like the genomes of all other living animals, is a collection of long polymers of DNA. These polymers are maintained in duplicate copy in the form of chromosomes in every human cell and encode in their sequence of constituent bases (guanine [G], adenine [A], thymine [T], and cytosine [C]) the details of the molecular and physical characteristics that form the corresponding organism. The sequence of these polymers, their organization and structure, and the chemical modifications they contain not only provide the machinery needed to express the information held within the genome but also provide the genome with the capability to replicate, repair, package, and otherwise maintain itself. In addition, the genome is essential for the survival of the human organism; without it no cell or tissue could live beyond a short period of time. For example, red blood cells (erythrocytes), which live for only about 120 days, and skin cells, which on average live for only about 17 days, must be renewed to maintain the viability of the human body, and it is within the genome that the fundamental information for the renewal of these cells, and many other types of cells, is found.

The human genome is not uniform. Excepting identical (monozygous) twins, no two humans on Earth share exactly the same genomic sequence. Further, the human genome is not static. Subtle and sometimes not so subtle changes arise with startling frequency. Some of these changes are neutral or even advantageous; these are passed from parent to child and eventually become commonplace in the population. Other changes may be detrimental, resulting in reduced survival or decreased fertility of those individuals who harbour them; these changes tend to be rare in the population. The genome of modern humans, therefore, is a record of the trials and successes of the generations that have come before. Reflected in the variation of the modern genome is the range of diversity that underlies what are typical traits of the human species. There is also evidence in the human genome of the continuing burden of detrimental variations that sometimes lead to disease.

Knowledge of the human genome provides an understanding of the origin of the human species, the relationships between subpopulations of humans, and the health tendencies or disease risks of individual humans. Indeed, in the past 20 years knowledge of the sequence and structure of the human genome has revolutionized many fields of study, including medicine, anthropology, and forensics. With technological advances that enable inexpensive and expanded access to genomic information, the amount of and the potential applications for the information that is extracted from the human genome is extraordinary.

Origins of the Human Genome

Comparisons of specific DNA sequences between humans and their closest living relative, the chimpanzee, reveal 99 percent identity, although the homology drops to 96 percent if insertions and deletions in the organization of those sequences are taken into account. This degree of sequence variation between humans and chimpanzees is only about 10-fold greater than that seen between two unrelated humans. From comparisons of the human genome with the genomes of other species, it is clear that the genome of modern humans shares common ancestry with the genomes of all other animals on the planet and that the modern human genome arose between 150,000 and 300,000 years ago.

Ongoing collaboration between archaeologists, anthropologists, and molecular geneticists at the Max Planck Institute in Germany and the Lawrence Berkeley National Laboratory and the Joint Genome Institute in the United States has enabled sequence comparisons between modern humans (Homo sapiens) and Neanderthals (H. neanderthalensis). The data obtained so far demonstrate that modern humans and Neanderthals share about 99.5 percent genome sequence identity; some scientists have claimed that sequence identity may actually be as high as 99.9 percent.

Research suggests that populations of H. sapiens split from H. neanderthalensis ancestral populations perhaps as recently as 370,000 years ago and likely shared a common ancestor some 500,000–700,000 years ago. Genomic studies have indicated that there was almost no interbreeding between H. sapiens and H. neanderthalensis. This suggests that when Neanderthals, the last of the Homo relatives of modern humans, became extinct about 30,000 years ago, only modern humans were left to populate Earth. However, other research has revealed that modern H. sapiens in Eurasia, specifically peoples in Europe, China, and Papua New Guinea, have genomes that are more similar to the Neanderthal genome than they are to the genomes of modern H. sapiens in Africa. Scientists estimate that 1 to 4 percent of DNA of modern Eurasians is shared with Neanderthals, a level of similarity that is not found between Neanderthals and modern Africans. These findings indicate that limited interbreeding and gene flow took place between Neanderthals and ancestral H. sapiens populations after the latter migrated out of Africa but before they dispersed to other parts of the world.

Comparing the DNA sequences of groups of modern humans from different continents also allows scientists to define the relationships and even the ages of these different populations. By combining these genetic data with archeological and linguistic information, anthropologists have been able to discern the origins of Homo sapiens in Africa and to track the timing and location of the waves of human migration out of Africa that led to the eventual spread of humans to other continents of the globe. For example, genetic evidence indicates that the first humans migrated out of Africa approximately 60,000 years ago, settling in southern Europe, the Middle East, southern Asia, and Australia. From there, subsequent and sequential migrations brought humans to northern Eurasia and across what was then a land bridge to North America and finally to South America.

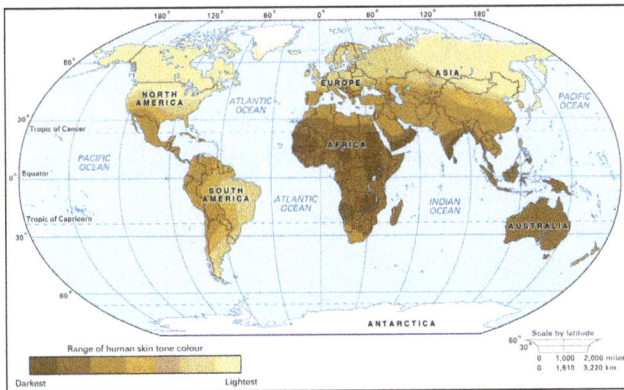

The global distribution of human skin colour is a well-defined example of genetic variation in which differential selective pressures favoured different characteristics in skin colour that conferred a survival advantage. Selective pressures for skin colour correlate with regional climate factors, such as latitude and sunlight. For example, the first populations of humans to settle in northern regions of the world were under selective pressure that favoured light skin colour to facilitate the absorption of sunlight, thereby preventing premature death from debilitating bone diseases.

As humans migrated across the continents, sequence variations arose that became differentially fixed in different populations. Some variations likely reflect what are called founder effects, changes in gene frequency that occur in small populations. Founder effects are generally characterized by genes that are expressed with increasing frequency from one generation to the next and can be traced back to the original founders of the population. Other variations reflect differential selective pressures at work. For example, populations living in equatorial climates were under strong selective pressure that favoured dark skin colour to protect against extreme sun exposure, thereby decreasing the deleterious health effects caused by sunburn and skin cancer. In contrast, populations migrating to more polar latitudes, where levels of sun exposure are relatively low, experienced strong selective pressure that favoured light skin colour, thereby facilitating the absorption of sunlight by the skin for the synthesis of vitamin D. In northern Europe and Scandinavia, therefore, individuals with genetic variations leading to lighter skin colour were less likely to become vitamin D deficient and suffer from the bone disease known as rickets.

Epigenome

The epigenome is a multitude of chemical compounds that can tell the genome what to do. The human genome is the complete assembly of DNA (deoxyribonucleic acid)- about 3 billion base pairs - that makes each individual unique. DNA holds the instructions for building the proteins that carry out a variety of functions in a cell. The epigenome is made up of chemical compounds and proteins that can attach to DNA and direct such actions as turning genes on or off, controlling the production of proteins in particular cells. When epigenomic compounds attach to DNA and modify its function, they are said to have "marked" the genome. These marks do not change the sequence of the DNA. Rather, they change the way cells use the DNA's instructions. The marks are sometimes passed on from cell to cell as cells divide. They also can be passed down from one generation to the next.

Functions of Epigenome

A human being has trillions of cells, specialized for different functions in muscles, bones and the brain, and each of these cells carries essentially the same genome in its nucleus. The differences among cells are determined by how and when different sets of genes are turned on or off in various kinds of cells. Specialized cells in the eye turn on genes that make proteins that can detect light, while specialized cells in red blood cells make proteins that carry oxygen from the air to the rest of the body. The epigenome controls many of these changes to the genome.

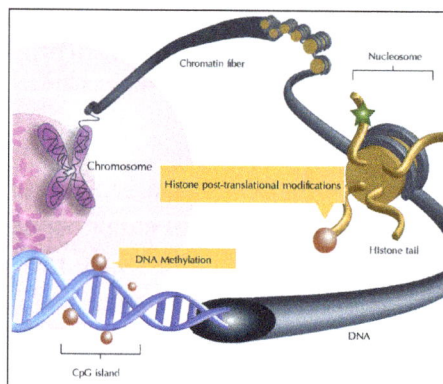

Structure of Epigenome

The epigenome is the set of chemical modifications to the DNA and DNA-associated proteins in the cell, which alter gene expression, and are heritable (via meiosis and mitosis). The modifications occur as a natural process of development and tissue differentiation, and can be altered in response to environmental exposures or disease.

The first type of mark, called DNA methylation, directly affects the DNA in a genome. In this process, proteins attach chemical tags called methyl groups to the bases of the DNA molecule in specific places. The methyl groups turn genes on or off by affecting interactions between the DNA and other proteins. In this way, cells can remember which genes are on or off.The second kind of mark, called histone modification, affects DNA indirectly. DNA in cells is wrapped around histone proteins, which form spool-like structures that enable DNA's very long molecules to be wound up neatly into chromosomes inside the cell nucleus. Proteins can attach a variety of chemical tags to histones. Other proteins in cells can detect these tags and determine whether that region of DNA should be used or ignored in that cell.

Inheritance

The genome is passed from parents to their offspring and from cells, when they divide, to their next generation. Much of the epigenome is reset when parents pass their genomes to their offspring; however, under some circumstances, some of the chemical tags on the DNA and histones of eggs and sperm may be passed on to the next generation. When cells divide, often much of the epigenome is passed on to the next generation of cells, helping the cells remain specialized.

Imprinting

The human genome contains two copies of every gene-one copy inherited from the mother and one from the father. For a small number of genes, only the copy from the mother gets switched on; for others, only the copy from the father is turned on. This pattern is called imprinting. The epigenome distinguishes between the two copies of an imprinted gene and determines which is switched on. Some diseases are caused by abnormal imprinting. They include Beckwith-Wiedemann syndrome, a disorder associated with body overgrowth and increased risk of cancer; Prader-Willi syndrome, associated with poor muscle tone and constant hunger, leading to obesity; and Angelman syndrome, which leads to intellectual disability, as well motion difficulties.

Changes in Epigenome

Although all cells in the body contain essentially the same genome, the DNA marked by chemical tags on the DNA and histones gets rearranged when cells become specialized. The epigenome can also change throughout a person's lifetime.

Lifestyle and environmental factors (such as smoking, diet and infectious disease) can expose a person to pressures that prompt chemical responses. These responses, in turn, often lead to changes in the epigenome, some of which can be damaging. However, the ability of the epigenome to adjust to the pressures of life appears to be required for normal human health. Some human diseases are caused by malfunctions in the proteins that "read" and "write" epigenomic marks.

Cancers are caused by changes in the genome, the epigenome, or both. Changes in the epigenome can switch on or off genes involved in cell growth or the immune response. These changes can lead to uncontrolled growth, a hallmark of cancer, or to a failure of the immune system to destroy tumors.

In a type of brain tumor called glioblastoma, doctors have had some success in treating patients with the drug temozolomide, which kills cancer cells by adding methyl groups to DNA. In some cases, methylation has a welcome secondary effect: It blocks a gene that counteracts temozolomide. Glioblastoma patients whose tumors have such methylated genes are far more likely to respond to temozolomide than those with unmethylated genes.

Changes in the epigenome also can activate growth-promoting genes in stomach cancer, colon cancer and the most common type of kidney cancer. In some other cancers, changes in the epigenome silence genes that normally serve to keep cell growth in check.

To compile a complete list of possible epigenomic changes that can lead to cancer, researchers in The Cancer Genome Atlas (TCGA) Network, which is supported by the National Institutes of Health (NIH), are comparing the genomes and epigenomes of normal cells with those of cancer cells. Among other things, they are looking for changes in the DNA sequence and changes in the number of methyl groups on the DNA.Understanding all the changes that turn a normal cell into a cancer cell will speed efforts to develop new and better ways of diagnosing, treating and preventing cancer.

Chromosomes

A chromosome is a string of DNA wrapped around associated proteins that give the connected nucleic acid bases a structure. During interphase of the cell cycle, the chromosome exists in a loose structure, so proteins can be translated from the DNA and the DNA can be replicated. During mitosis and meiosis, the chromosome becomes condensed, to be organized and separated. The substance consisting of all the chromosomes in a cell and all their associated proteins is known as chromatin. In prokaryotes, there is usually only a single chromosome, which exists in a ring-like or linear shape. The chromatin of most eukaryotic organisms consists of multiple chromosomes, Each chromosome carries part of the genetic code necessary to produce an organism.

Having the entire genetic code divided into different chromosomes allows the possibility of variation through the different combinations of chromosomes with the different alleles, or genetic variations that they contain. The recombination and mutation of chromosomes can occur during mitosis, meiosis, or during interphase. The end result is organisms that function and behave in different ways. This variation allows populations to evolve over time, in response to changing environments.

Function of a Chromosome

The chromosome holds not only the genetic code, but many of the proteins responsible for helping express it. Its complex form and structure dictate how often genes can be translated into proteins, and which genes are translated. This process is known as gene expression and is responsible for creating organisms. Depending on how densely packed the chromosome is at certain point determines how often a gene gets expressed. As seen in the image of chromosome structure shown below, less active genes will be more tightly packed than genes undergoing active transcription. Cellular molecules that regulate genes and transcription often work by activing or deactivating these proteins, which can contract or expand the chromosome. During cell division, all the proteins are activated and the chromatin becomes densely packed into distinct chromosomes. These dense molecules have a better chance of withstanding the pulling forces that occur when chromosomes are separated into new cells.

Chromosome Structure

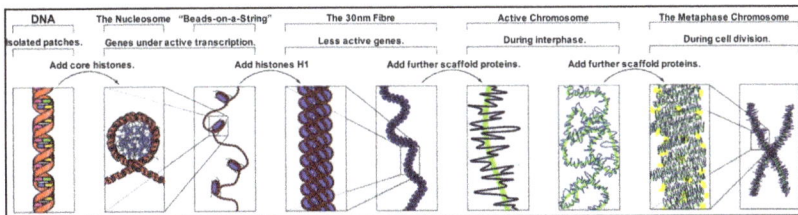

As seen in the graphic above, chromosomes have a very complex structure. DNA, or deoxyribonucleic acid makes the base of the structure, as seen on the far left. DNA is made of a two strings of nucleic acid base pairs. The base pairs in DNA are cytosine, adenine, thymine, and guanine. The spiral structure formed by the two strings of DNA is due to complimentary pairing between every base with its pair on the opposite string. Adenine pairs with thymine and guanine pairs with cytosine. The opposite side of the bases form a phosphate-deoxyribose backbone, which keeps the strands intact.

When the DNA is duplicated, the strands are separated, and a polymerase molecule builds a new string that corresponds to each side. In this way, the DNA is perfectly replicated. This can be done artificially by a polymerase chain reaction in which special enzymes and heat are used to separate and replicated the strings a number of times, to produce many copies of the same DNA. This makes it much easier to study any string of DNA, even whole chromosomes or genomes.

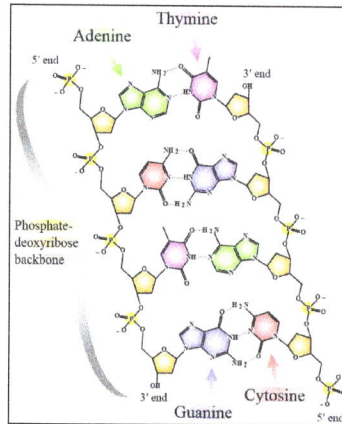

DNA chemical structure.

After the cell has expressed and duplicated the DNA, cell division can occur. This occurs in both prokaryotes and eukaryotes, but only eukaryotes condense their DNA so it can be separated. Prokaryotic DNA is so simple that relatively few structural proteins are associated with the chromosome. In eukaryotes, many structural proteins are used.

The first of these proteins are core histones. Many individual histone proteins bind together to form a core histone. The DNA can wrap around one of these histones, giving it a wound structure. This structure, and the associated histone, is known as the nucleosome. These nucleosomes form "beads-on-a-string". The string becomes wound back and forth by another histone, histone H1, and eventually fibers are produced. The next type of protein, scaffold proteins, start to wind the fiber into a loose structure. When the chromosome must condense during cell division, more scaffold proteins are activated, and the structure becomes much denser. In fact, even with a microscope, individual chromosomes cannot be discerned until near the middle of cell division cycles, when the chromosome becomes very dense. This process is seen as the pictures progress towards the right.

4

Cell-Cell Interactions

The direct interactions between cell surfaces are known as cell-cell interactions. They play an important role in the function and development of multicellular organisms. The defense system of the host which protects it from disease is known as the immune system. This chapter has been carefully written to provide an easy understanding of the varied facets of cell-cell interaction such as cell signalling and cell-cell adhesion, and biology of the immune system.

Cell Signalling

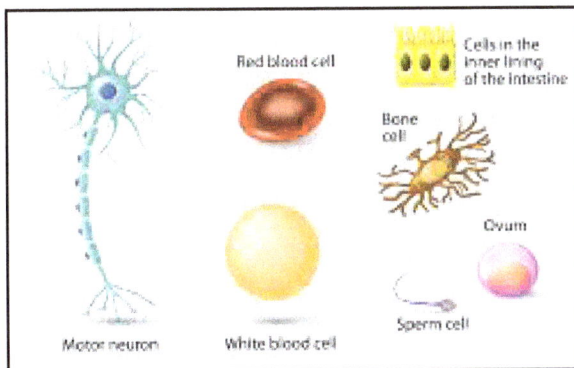

Essential to the survival of every cell is to monitor the environment and to respond to external stimuli. For most cells this includes appropriate communication with neighboring cells. Cell signaling (or signal transduction) involves:

- Detection of the stimulus (in most cases a molecule secreted by another cell) on the surface of the plasma membrane.

- Transfer of the signal to the cytoplasmic side.

- Transmission of the signal to effector molecules and down a signaling pathway where every protein typically changes the conformation of the next down the path, most commonly by phosphorylation (by kinases) or dephosphorylation (by phosphatases).

The final effect is to trigger a cell's response, such as the activation of gene transcription.

Types of Signaling

Cells communicate by means of extracellular signaling molecules that are produced and released by signaling cells. These molecules recognize and bind to receptors on the surface of target cells where they cause a cellular response by means of a signal transduction pathway.

Depending on the distance that the signaling molecule has to travel, we can talk about three types of signaling:

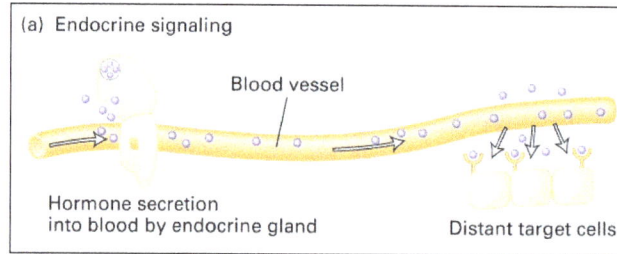

(a) Endocrine signaling

Blood vessel

Hormone secretion into blood by endocrine gland

Distant target cells

In endocrine signaling hormones are produce by an endocrine gland and sent through the blood stream to distant cells.

Hormones can be: Small lipophilic molecules that diffuse through the cell membrane to reach cytosolic or nuclear receptors. Examples are progesterone and testosterone, as well as thyroid hormones. They generally regulate transcription; or water soluble molecules that bind to receptors on the plasma membrane. They are either proteins like insulin and glucagons, or small, charged molecules like histamine and epinephrine.

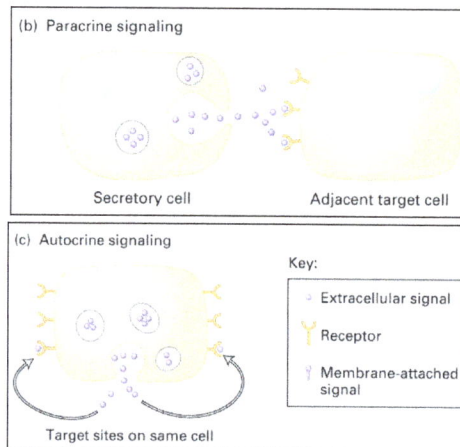

(b) Paracrine signaling

Secretory cell

Adjacent target cell

(c) Autocrine signaling

Key:

Extracellular signal

Receptor

Membrane-attached signal

Target sites on same cell

In paracrine signaling the signaling molecule affects only target cells in the proximity of the signaling cell. An example is the conduction of an electric signal from one nerve cell to another or to a muscle cell. In this case the signaling molecule is a neurotransmitter.

In autocrine signaling cells respond to molecules they produce themselves. Examples include many growth factors. Prostaglandines, lipophilic hormones that bind to membrane receptors, are often used in paracrine and autocrine signaling. They generally modulate the effect of other hormones.

Once a signaling molecule binds to its receptor it causes a conformational change in it that results in a cellular response. The same ligand can bind to different receptors causing different responses (e.g.. acetylcholine). On the other hand, different ligands binding to different receptors can produce the same cellular response (e.g. glucagon, epinephrine).

Types of Receptors

There are a number of receptor classes that are used in different signaling pathways. The two more predominant are:

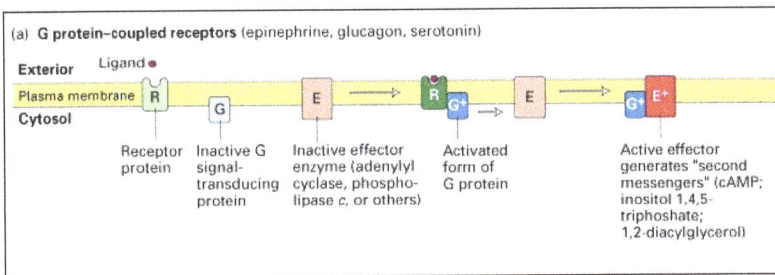

(a) **G protein–coupled receptors** (epinephrine, glucagon, serotonin)

The conformational change in the receptor upon ligand binding activates a G protein, which in turns activates an effector protein that generates a second messenger.

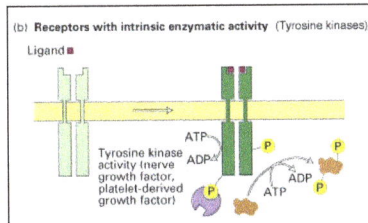

(b) **Receptors with intrinsic enzymatic activity** (Tyrosine kinases)

These receptors have a catalytic activity that is activated by binding of the ligand. An example are tyrosine-kinase receptors. Binding of an often dimeric ligand induces dimerization of the receptors that leads to cross phosphorylation of the cytosolic domains and phosphorylation of other proteins.

Identification and Purification of Cell-Surface Receptors

Hormone receptors are difficult to identify and purify because they are present in very low abundance and they have to be solubilized with nonionic detergents. Given their high specificity and high affinity for their ligands, the presence of a certain receptor in a cell can be detected and quantified by their binding to radioactivelylabeled hormones. The binding of the hormone to a cell suspensions increases with hormone concentration until it reaches receptor saturation. Specific binding is obtained by measuring both the total and the non-specific binding (which is obtained by using a large excess on unlabeled hormone).

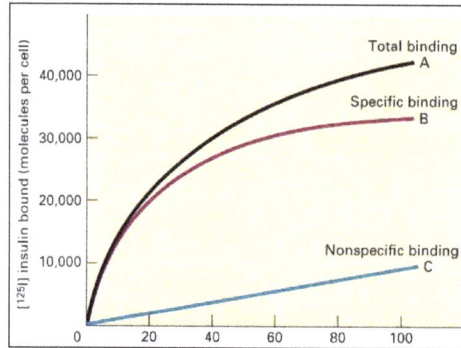

The receptor can be purified in some cases by means of affinity chromatography: the hormone is linked to agarose beads. Crude, solubulized membranes are passed through a column with these beads, which will retain only their specific receptor. The receptor is later release by passing excess hormone through the column. A single pass can produce a 100,00 fold enrichment.

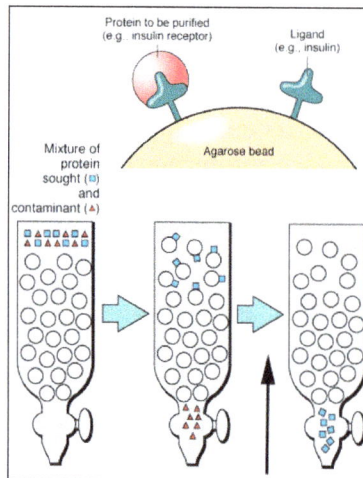

In many cases, however, the number of receptor on the cell surface is to low to use chromatography. An alternative is to produce recombinant receptor protein.

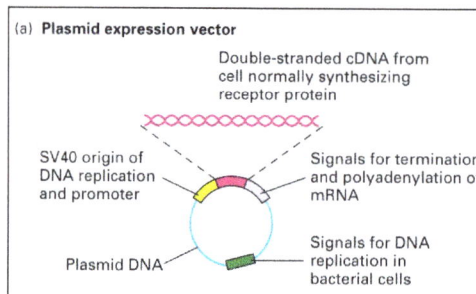

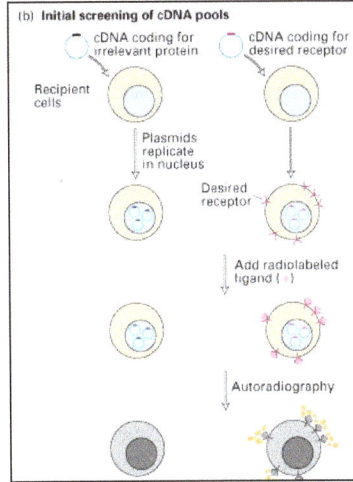

(b) Initial screening of cDNA pools

A plasmid cDNA library from the cells expressing the receptor is screen by transfecting the cloned cDNAs into cells that normally do not express the receptor. Cells that take up the cDNA encoding the receptor are detected by their binding to fluorescence or radioactively-labeled hormone. The cDNA can then be sequenced and used produce large amounts of recombinant protein.

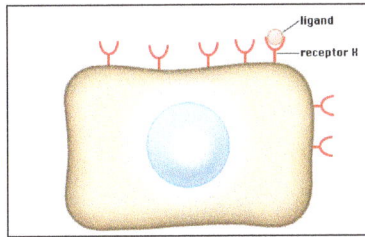

Other Conserved Functions

The three main classes of intracellular signaling proteins are:

G proteins (GTPase switch proteins) - These proteins change between an active conformation when bound to GTP, and an inactive conformation when bound to GDP. IN the absence of a signal they are bound to GDP. Signal results in the release of GDP and the binding of abundant GTP. After a short period of time they hydrolyse GTP and come back to their "off" state.

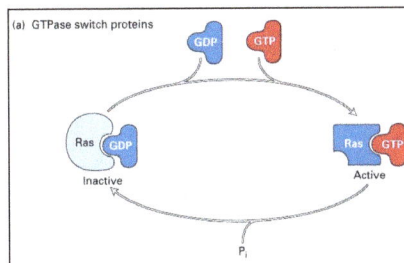

(a) GTPase switch proteins

Protein Kinases - Upon activation they add phosphate groups to themselves and/or other proteins at either serine/threonine, or at tyrosine residues. Their activity can be regulated by second messengers, interaction with other proteins, or by phosphorylation itself. They are opposed by phosphatases that remove phosphate groups from specific phosphorylated proteins.

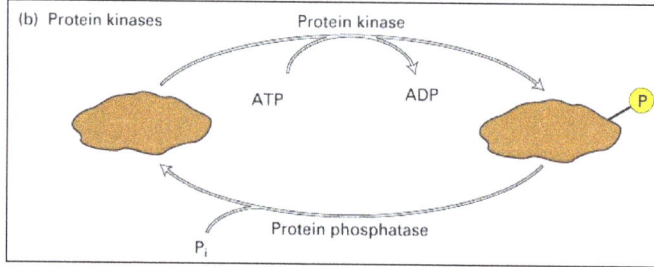

Adaptor Proteins – Many signaling pathways require the formation of large protein complexes that are held together by adaptor proteins. These proteins contain several specialized domains that act as docking sites for other proteins (e.g. SH2 domains bind to phosphotyrosines; SH3 domains bind to proline-rich sequences).

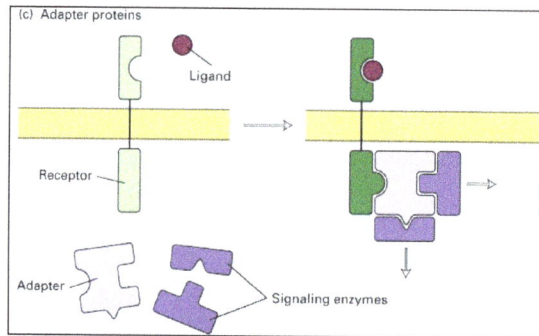

G Proteins and G Protein-Coupled Receptors

Heterotrimeric G proteins

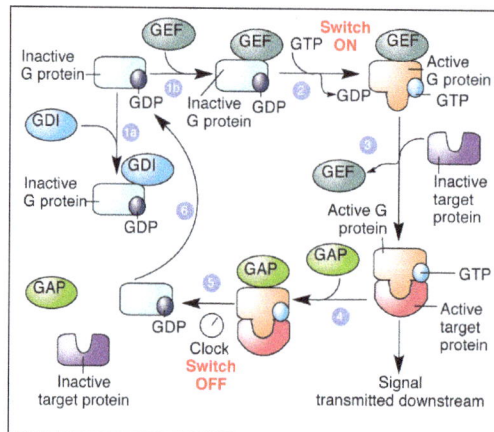

GTP-binding proteins (or G-proteins) are molecular switches that turn activities on and off. In their GTP-bound state they are active and bind to specific proteins downstream, while in their GDP state they are inactive.

Associated proteins include GTPases-activating proteins (GAPs), guanine nucleotide-exchange factors (GEFs) and guanine nucleotide-dissociation inhibitors (GDIs).

G protein-coupled Receptors

- Many different mammalian cell-surface receptors are coupled to a heterotrimeric signal-transducing G protein, covalently linked to a lipid in the membrane.

- Ligand binding activates the receptor, which activates the G protein, which activates an effector enzyme to generate an intracellular second messenger.

- All G protein-coupled receptors (GPCRs) contain 7 membrane-spanning regions with their N-terminus on the exoplasmic face and C-terminus on the cytosolic face. They are ligand specific and differ in their extracellular surface.

- GPCRs are involved in a range of signaling pathways, including light detection, odorant detection, and detection of certain hormones and neurotransmitters.

- Seven-helix transmembrane receptors and heterotrimeric G proteins exist in different isoforms: epinephrine binding to -adrenergic receptors in cardiac muscle activates a Gs that stimulates cAMP production, inducing muscle contraction. Its binding to the - adrenergic receptor in intestine activates a Gi that inhibits cAMP inducing muscle relaxation.

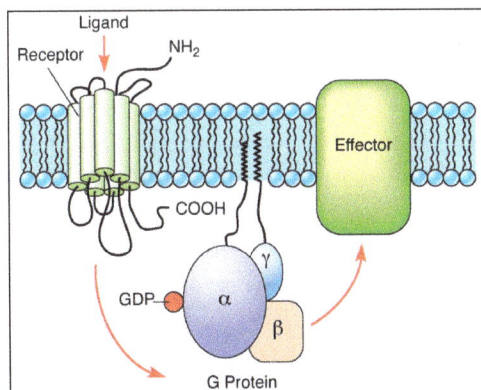

Activation of the G protein by the Receptor

Ligand binding changes the conformation of the receptor so that it binds the G protein. This in turn causes the exchange of GDP for GTP in the subunit, switching it to the activated state. While a ligand is bound it can activate many G protein molecules.

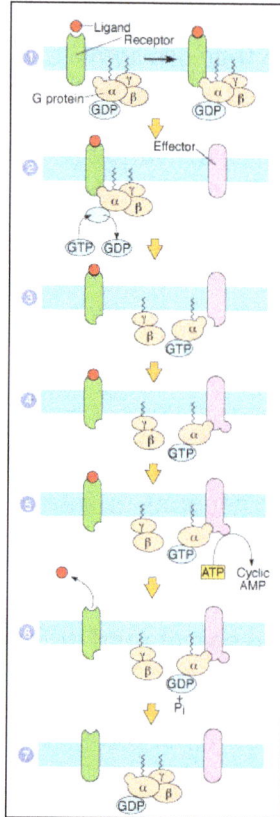

Relay of the Signal to the Effector

With bound GTP $G\alpha$ dissociates from the $G\beta\gamma$ complex and is able to bind to its effector molecule. Binding of G causes the activation of the effector. G can in turn also activate downstream effectors.

Ending the response. When GTP is hydrolyzed the $G\alpha$ subunit rebinds to $G\beta\gamma$, returning the complex to its inactivated state. Although the $G\alpha$ generally has a low GTPases activity, this can be accelerated by interaction with its GAP protein.

Deactivation by phosphorylation and binding of arrestin.

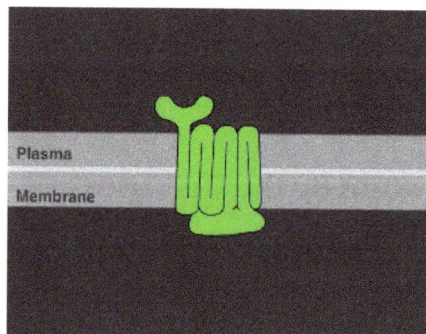

Second Messenger cAMP

A second messenger is a substance that is released in the cytoplasm following activation of a receptor. It is non-specific and can generate a variety of responses in the cell. cAMP is an example.

cAMP is synthesized by an integral membrane protein, adenylyl cyclase, using ATP as a substrate.

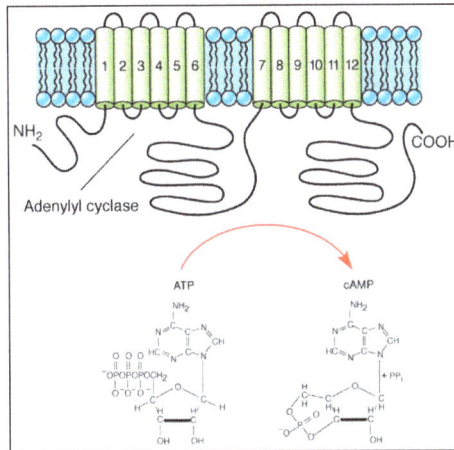

Adenylyl cyclase can be activated or inactivated by different G proteins following the binding of an inhibitory or stimulatory ligand to G protein coupled receptors.

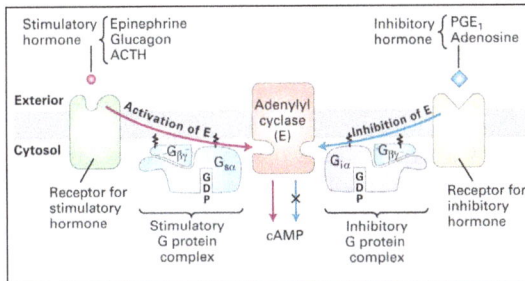

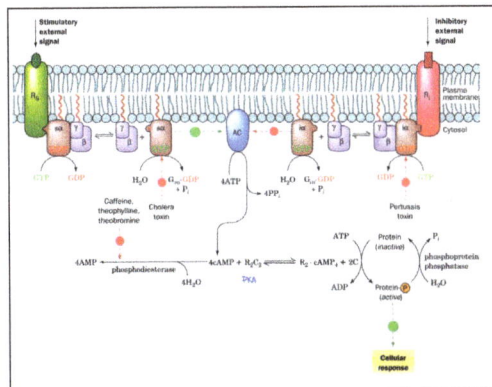

Glucose Mobilization

- Glucose is stored in animal cells as glycogen, an insoluble polymer.

- The enzyme phosphorylase catalyzes glycogen breakdown, while glycogen synthase catalyzes polymerization.

- Regulation is achieved by several hormones, such as glucagon (from the pancreas) and epinephrine (from the adrenal medulla).

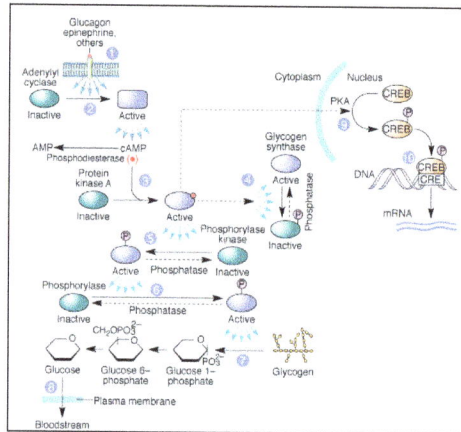

- Binding of epinephrine or glucagon to their receptors activates adenyl cyclase.

- cAMP binds to the regulatory subunit of protein kinase A (PKA) causing the release of the catalytic subunit.

- PKA phosphorylates and inhibits glycogen synthase.

- PKA phosphorylates and activates phosphorylase kinase, which then phosphorylates and activates phosphorylase.

- PKA also translocates to the nucleus where it phosphorylates the transcription factor CREB. Phosphoylated CREB binds to the CRE enhancer activating genes involved in gluconeogenesis.

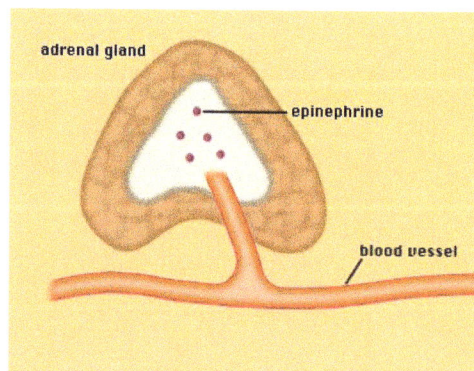

Amplification

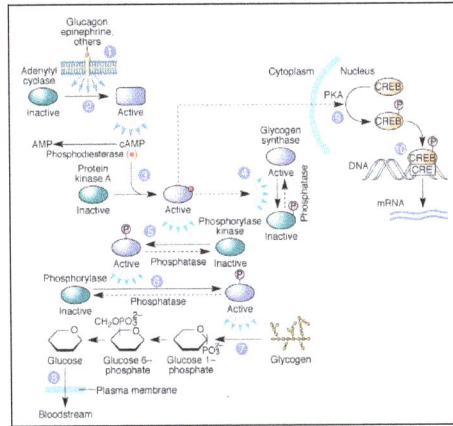

Because the concentration of hormones in the blood is very low (<10⁻⁸ M), the signal has to be amplified:

- Binding of a single ligand stimulates a large number of adenylyl cyclases,

- Each producinga large number of cAMPs,

- Each two cAMPs activates a PKA,

- Which in term phosphorylates a number,

- Of other proteins, and so on.

Termination

Termination of the signal involves phosphatase-1, which removes phosphates groups from the different enzymes. Its activity is regulated by inhibitor-1, which itself is activated by PKA. Because cAMP is continually degraded., when the hormone dissociates from its receptor and adenylyl cyclase is shut down, cAMP levels quickly drop. This inactivates PKA and thus inhibitor-1, leading to the activation of phosphatase-1.

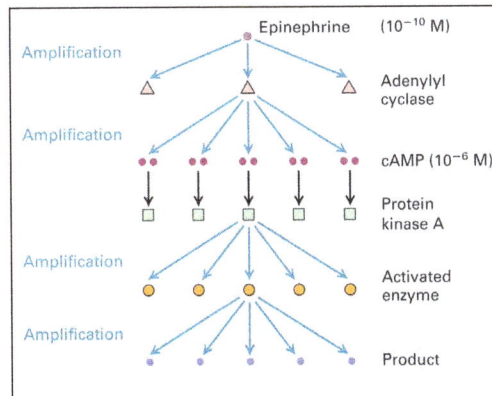

The figure below shows the experimental demonstration that B- adrenergic receptors mediate the induction of cAMP by epinephrine.

1. Target cells lacking the -adrenergic receptor but containing Adenylyl cyclase and G proteins fail to respond to epinephrine and their cytosolic cAMP remains low.

2. Purified liposomes with the receptor are incubated with these cells and fuse, incorporating the receptor to the cell surface.

3. When exposed to epinephrine these cells now quickly increase their cytosolic cAMP levels.

Wild type and chimeric α - and β -adrenergic receptors were expressed in Xenopus oocytes. Agonist ligands that bind selectively to one or the other where used to determined the regions of the receptor involved in ligand binding and in interaction with the G protein (Gi or Gs).

- Chimera 1 acts like a β -adrenergic receptor.

- Chimera 2 bindings the α agonist, but still activates cAMP (binds Gs).

- Chimera 3 binds the β agonist but inhibits cAMP (binds Gi).

In conclusion, the C-terminal fragment is involved in ligand binding/recognition, while regions 5 and 6 interact with the G protein.

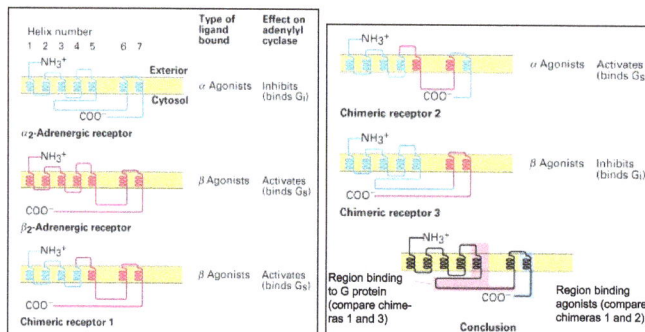

Lipid-derived Second Messengers

Activated phospholipases (hydrolytic enzymes that split phospholipids) can convert certain cell membrane phospholipids into second messengers.

Phosphatidylinositol (PI)

- PI can be phosphorylated to PIP and to PIP_2 by different kinases that are themselves activated by G protein couple receptors (GPCRs) and receptor tyrosine kinases.

- PIP$_2$ can by cleaved by phopholipase C (PLC), which is itself activated, depending on the isoform, by GPCRs and RTKs. The two products of the cleavage, the lipophilic DAG, and the soluble IP$_3$, are secondary messengers in several signaling pathways.

- IP$_3$ binds to a IP$_3$-gated Ca^{++} channel in the ER membrane that opens upon IP3 binding increasing cytosolic Ca^{++} concentrations.

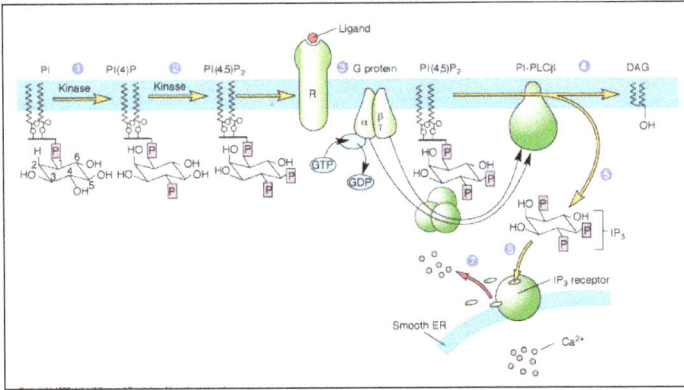

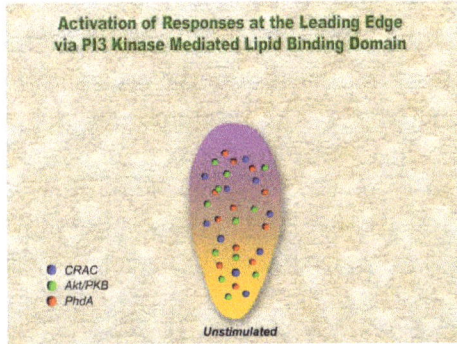

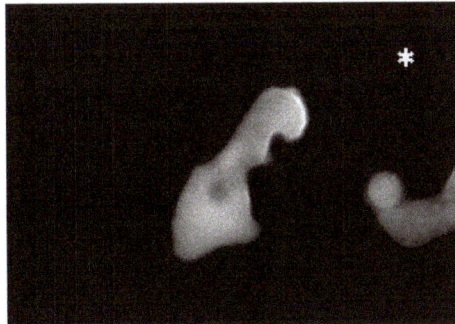

Activation of Responses at the Leading Edge via PI3 Kinase Mediated Lipid Binding Domain

Calcium Signaling

The concentration of Ca^{++} is kept very low in the cytosol ($10^{-7} - 10^{-8}$ M), 10,000 times lower than in the extracellular space, by means of Ca++ pumps that pump these ions out of the cell or into the ER.

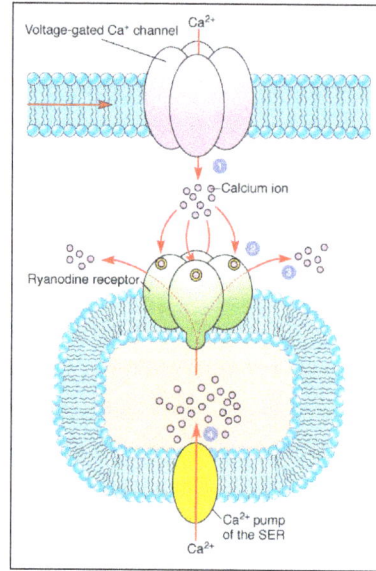

Regulated Ca++ channels open upon different stimulae to transiently let Ca++ into the cytosol where it serves as a secondary messenger in several signaling pathways.

The main types of Ca++ channels are the IP$_3$-receptor discussed above and the ryanodine receptor. The latter exists in the ER of nerve cells and the sarcoplasmic reticulum of muscle cells where, upon arrival of the action potential, opens to trigger calcium influx into the cytosol that results in muscle contraction. Ryonodine receptors are always in close proximity to voltage-gated Ca++ channels and other voltage-sensitive receptors in the plasma membrane that are involved in their activation.

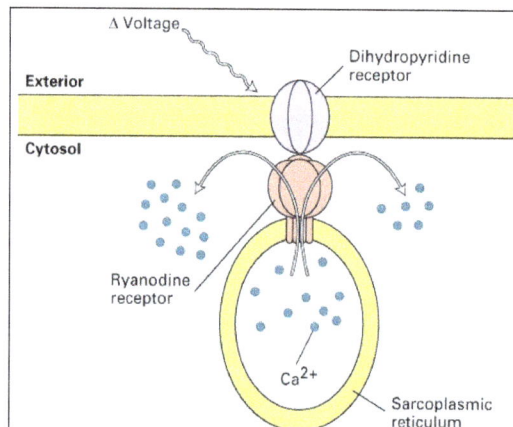

Ca++ is involved in a number of signaling pathways. Elevation of Ca++ levels via the inositollipid signaling pathway involves protein kinase C (PKC). Binding of Ca++ to the kinase recruits it to the membrane where its kinase activate is stimulated by interaction with DAG. PKC then phosphorylates a number of substrates further along the pathways.

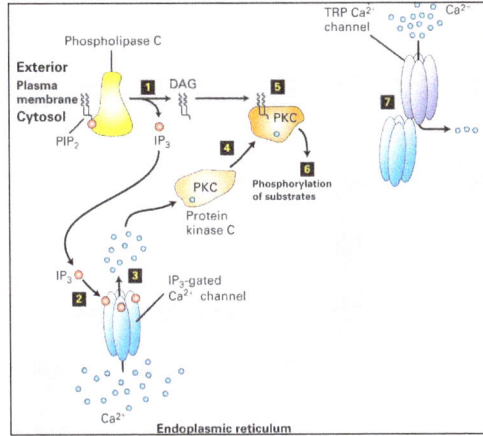

Ca^{++} effects includes exocytosis of secretory vesicles, muscle contraction or the inducement of mitosis in fertilized eggs.

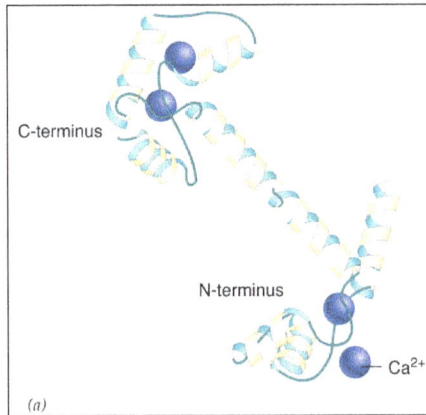

(a)

To trigger these responses calcium affects a number of cellular effectors. In most cases it does it in conjunction with the calcium-binding protein calmodulin. This protein is found in all eukaryotes and is widely conserved in sequence among species.

When the concentration of Ca^{++} increases calmodulin binds 4 Ca^{++} ions. This results in a large conformational change in the protein that increases its affinity for a number of effector proteins.

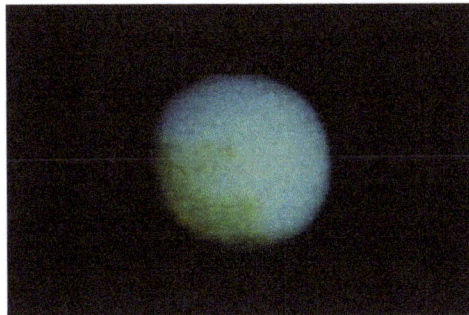

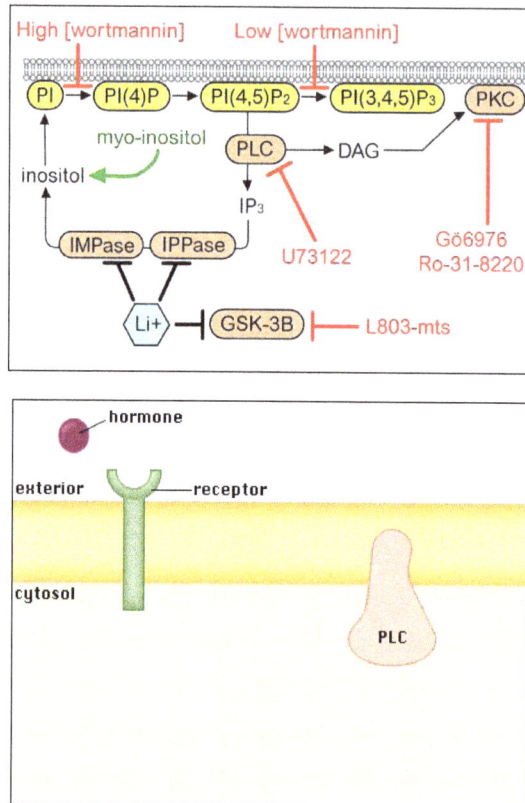

Receptor Tyrosine Kinases

Insulin Signaling

- Insulin signals the removal of glucose from the blood and synthesis of glycogen.

- The insulin receptor is a member of the tyrosine kinase superfamily, generally involved in cell growth and differentiation.

- The insulin receptor tyrosine kinase (RTK) is a tetramer of two extracellular α subunits and two transmembrane β subunits with a single transmembrane region.

- Subunits bind insulin and induce a conformational change in the β subunits, activating their cytoplasmic kinase domain.

- Activated β subunits phosphorylate one another and a variety of insulin receptor substrates (IRSs).

- RTKs phosphorylate only phosphotyrosine motifs. After phosphorylation these motifs have a high affinity for SH2 domains.

- Interaction of phosphotyrosine motifs with SH2 proteins cause their conformational change so that they bind to other proteins or are translocated to other parts of the cell.

- Different SH2 proteins will activate separate signaling pathways.

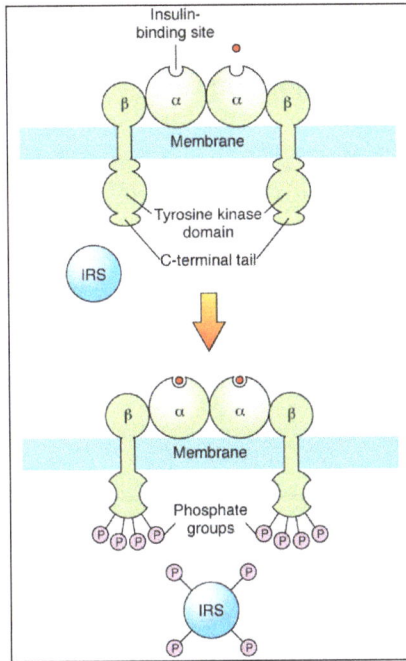

Insulin Stimulation Activates Protein Kinase B and MAP Kinase

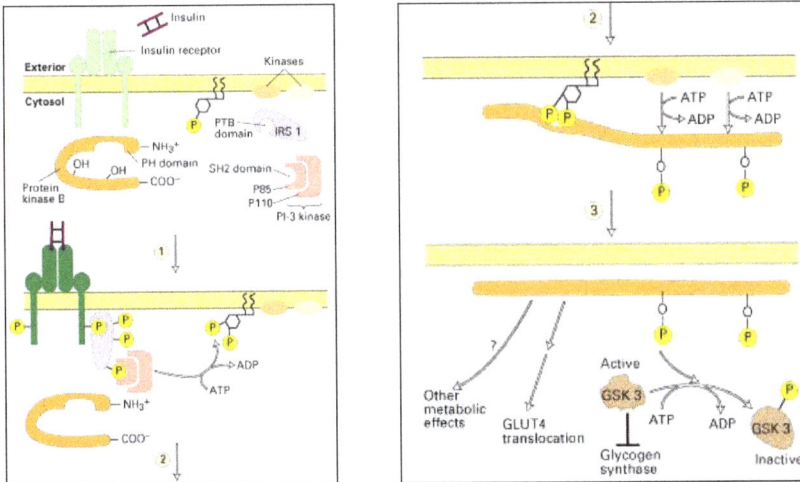

Activation of insulin RTKs results in:

- Transfer of glucose transporters to the plasma membrane.

- Increase in protein synthesis.

- Stimulation of Glycogen synthases.

- Activation of phosphatase 1.

General RTKs, Ras and the MAP Kinase Cascade

RTKs are receptors for a variety of extracellular ligands:

- Hormones: insulin, growth hormone.

- Growth factors like EGP.

- Cytokines (secreted by certain immune cells to affect others: interferones, interleukins).

- Most RTKs are monomers that only dimerize when bound to their ligand.

- Dimerization activates the kinase activity and leads to autophosphorylation, creating sites for interactions with specific effectors. A key component of the RTK cascade is Ras.

- Ras

  ○ Ras is small, monomeric G protein with very low GTPase activity on its own. GAPs activate hydrolysis in Ras by 105 fold.

  ○ Ras is mutated in 30% of human tumors. Most mutations block hydrolysis keeping Ras active and the cell in a proliferative state.

- Grb2

  ○ SH2 protein that binds to phosphorylated RTKs.

  ○ It has three domains, one binds the RTK, the other binds Sos.

- Sos

  ○ It is a Ras-GEF. When recruited to the membrane it activates Ras.

- Raf

  ○ Ras-GTP recruits Raf, which becomes activated as a protein kinase and initiates the MAP kinase cascade.

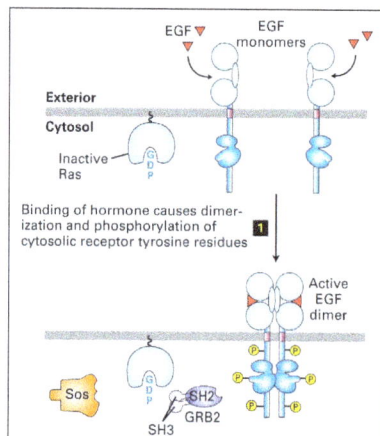

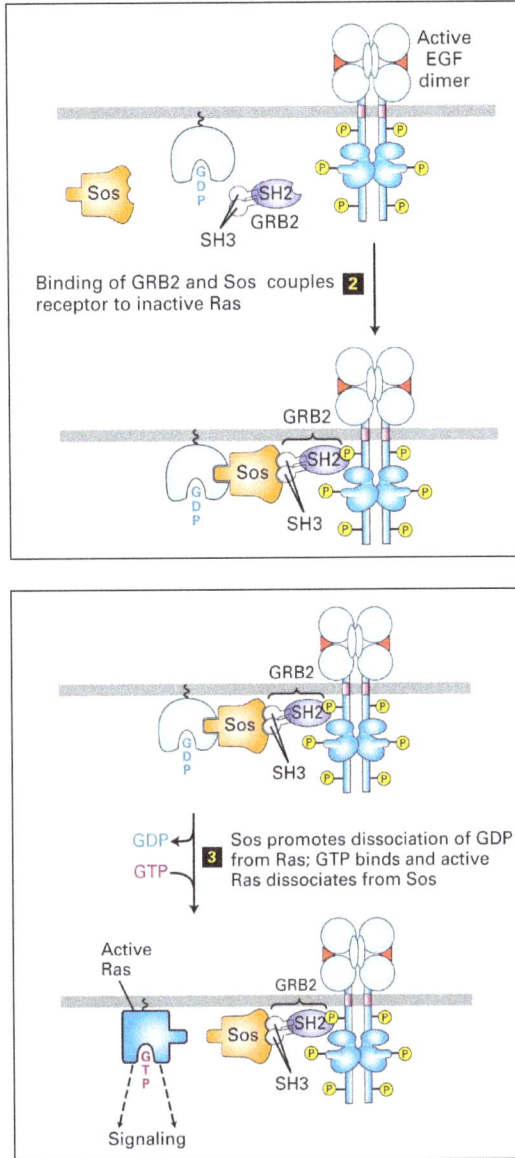

Ras

The experimental support for placing Ras down stream from RTKs in the signaling pathway was obtained in studies of cultured fibroblasm responding to growth factors. When antibodies against Ras were microinjected into these cells they failed to proliferate in the presence of growth factors. Microinjection of a constitutively active Ras mutant lacking GTPase activity resulted in proliferation, even in the absence of growth factors.

Additional studies show that addition of growth factors leads to a rapid increase in the proportion of Ras that is bound to GTP.

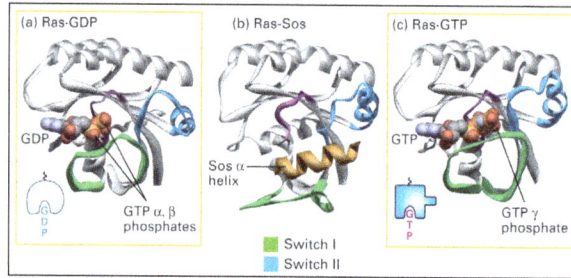

(a) Ras-GDP (b) Ras-Sos (c) Ras-GTP

Both biochemical and X-ray crystallographic studies have shed light on the mechanism by which Sos acts as an exchange factor for Ras. Binding of Sos to Ras changes the conformation of this protein and opens the nucleotide binding site, resulting in GDP release (a). Binding of the abundant GTP changes the conformation of Ras and displaces Sos (b). The switch I and II regions are those most affected in conformation by the nucleotide state.

MAP Kinase Cascade

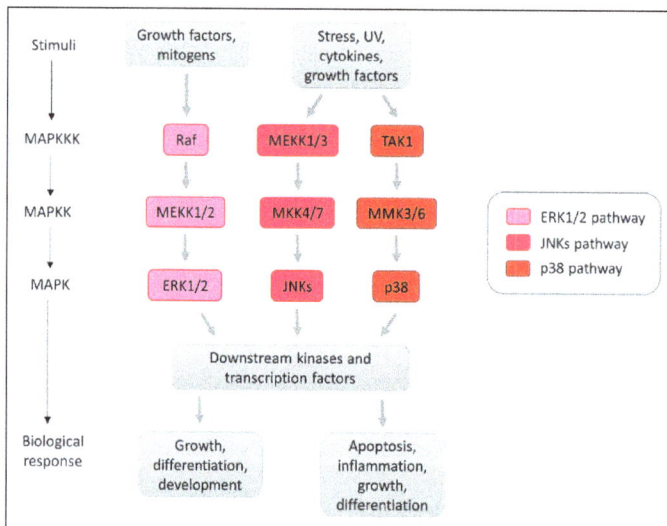

(Mitogen-Activated Protein: activated by a mitosis-stimulating growth factor).

- Recruitment of Raf by activated Ras to the plasma membrane activates it as a kinase (MAPKKK).

- Raf phosphorylates MAPKK.

- MAPKK phosphorylates MAPK.

- MAPK phosphorylates transcription factors like Elk-1.

- Elk-1 activates the transcription of Fos and June. These proteins then make a dimer that activates the transcription of cell proliferating genes.

- The transcription of MAPK phosphatase (MKP-1) is also activated. MKP-1 dephosphorylates MAPK and stop signalling.

- This pathway is generally used in eukaryotes for many different functions. Different types of information are transmitted thanks to the existence of different isoforms for each of the cascade proteins.

The human genome encodes for 2000 kinases and 1000 phosphatases.

MAP Kinase

In its unphosphorylated state MAPK is inactive, its catalytic site blocked by the "phosphorylation lip" (a). Binding of MAPKK (MEK) to MAPK and phosphorylation of a Tyr and a Thr in the phosphorylation lip alters the conformation of the lip, allowing the binding of ATP to the catalytic site (b).

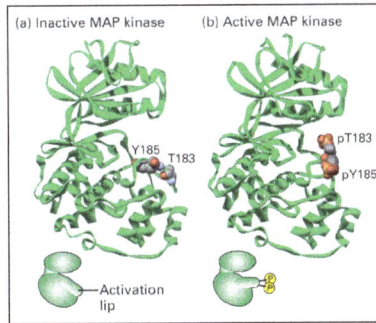

The phosphotyrosine also contributes to the binding of MAPK substrates. In addition, the phosphorylation of the two residues in the lip region promotes dimerization of MAPK. This dimeric form can translocate into the nucleus where it regulates the activity of transcription factors.

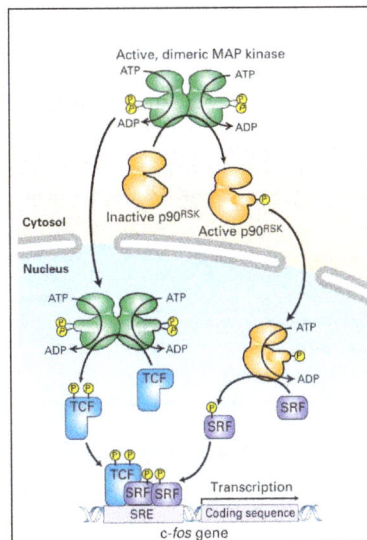

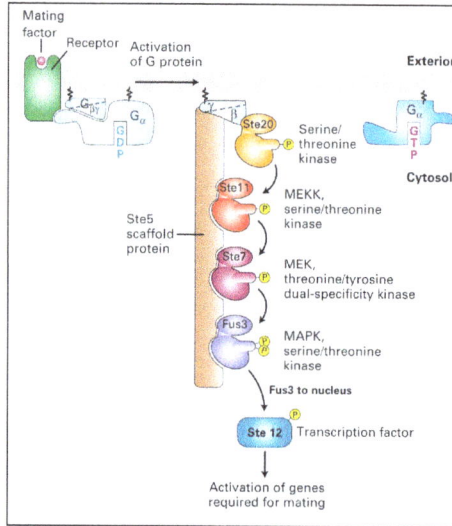

Integrin Signaling

- Normal cells cannot grow in suspension (as cancer cells do), but need to attach to a surface that resembles the extracellular matrix (ECM) in a tissue.

- Cells have transmembrane proteins called integrins that anchor them to materials in the ECM (fibronectin, collagen, proteoglycans, etc).

- Interaction of integrins with an extracellular ligand generates a variety of signals, some essential for cell growth and differentiation.

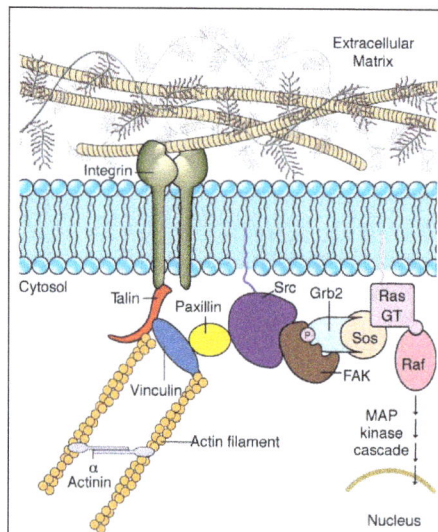

The interaction of the cell with the ECM is done through focal adhesions containing clustered integrins, cytoplasmic proteins and actin stress fibers.

The tyrosine kinase Src is localized to focal adhesions. Src phosphorylates the focal adhesion kinase (FAK).

Phosphorylated FAK is bound by SH2 proteins like Grb2-Sos, activating the MAP kinase cascade.

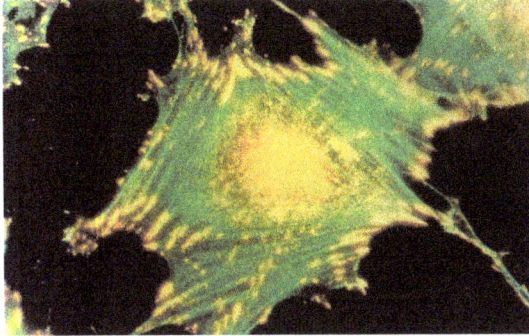

Activation of FAK also results in signals that affect ribosomal components resulting in the regulation of protein synthesis required for the transition from G1 to S phase.

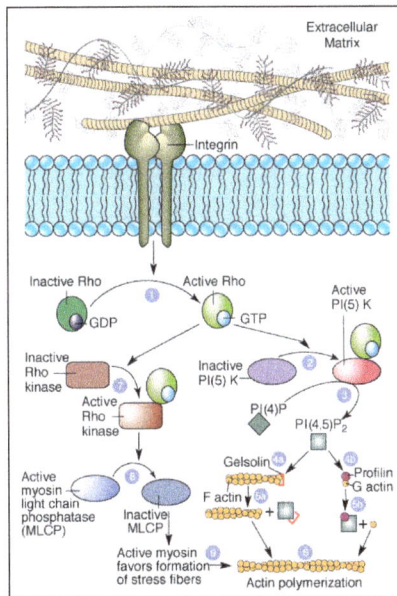

The assembly of focal adhesions has a large effect in the organization of the actin cytoskeleton in the cell through the G protein Rho.

On one pathway activated Rho activates PIP2 signaling that affects actin binding proteins controlling polymerization.

In another pathway Rho activates the Rho kinase that inactivates a myosin light chain phosphatase, leading to the activation of myosin and the organization of actin into stress fibers.

Relationship between Signaling Pathways

A single cell has dozens of different receptors sending signals to the inside of the cell simultaneously. These signals, along with different pathways, are integrated in the cell to give rise to a coordinated response. Signaling pathways can relate to each other in different ways:

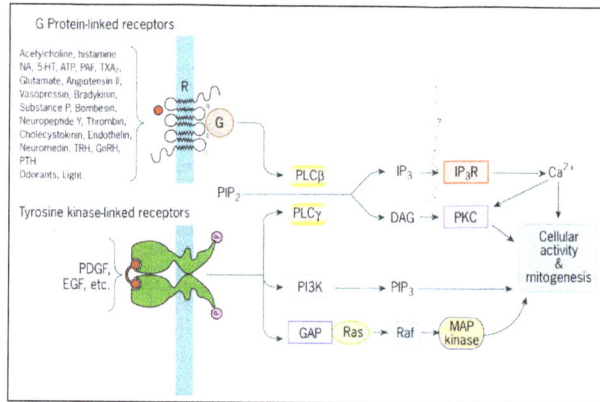

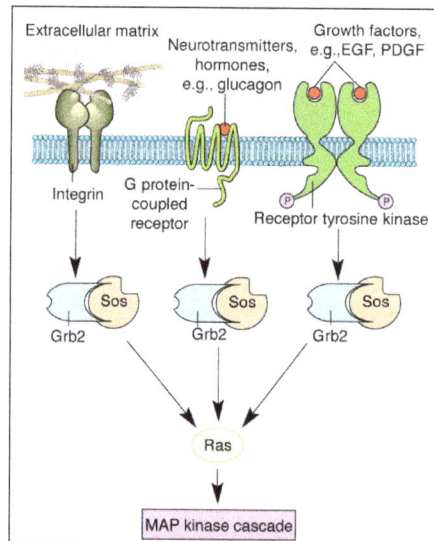

Converging Signaling Pathways

G protein-coupled receptors, RTKs and integrins can all relay signals that result in the recruitment to the membrane of the adaptor protein Grb2 and the activation of Ras and the MAP kinase cascade.

Diverging Signaling Pathways

Activation of the EGF RTK generates phosphotyrosine sites that interact with a variety of SH2 proteins, thus sending a signal down several pathways.

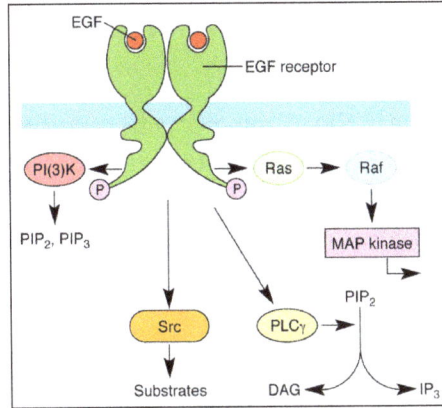

Crosstalk between Signaling Pathways

EGF/epinephrin.

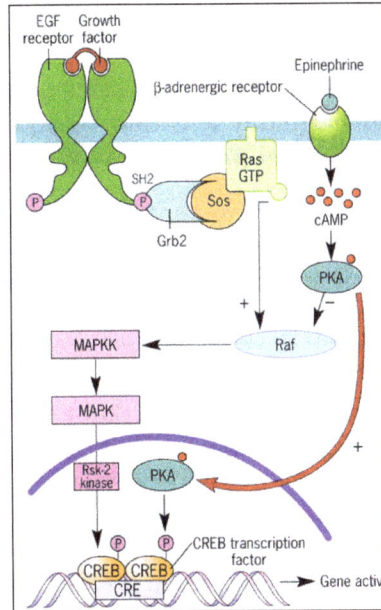

Hormone Responses in Glucose Metabolism: Liver and Muscle Cells

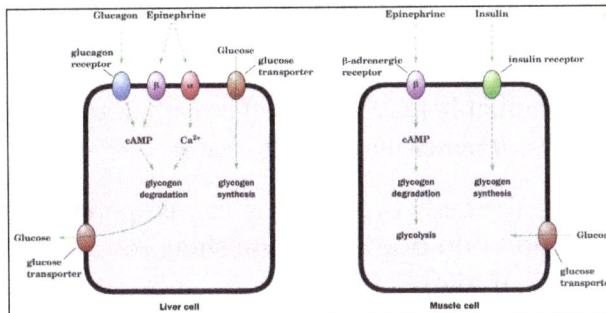

Cell Proliferation

Cell proliferation is an increase in the number of cells resulting from the normal, healthy process by which cells grow and divide. In this regard, cell proliferation can be a good indicator of general cell health. Cells that are subject to a variety of disease states may exhibit different rates of proliferation than normal cells. Therefore, measuring rates of cell proliferation between different cell populations may provide insight into the relative health of those cells.

There are several methods available to measure cell proliferation rates. One method is to measure the overall metabolic activity inside a cell. Several dyes are available that can permeabilize a cell and react with certain enzymes and other factors and form a colored end-product which can be easily detected. One such dye, MTT, is a well-published molecule that produces a distinct purple color when added to proliferating cells. Another dye, WST-1, provides a similar color change in proliferating cells, but has the advantage of a single-step method that eliminates the detergent solubilization step required for MTT.

Fluorescence dyes are also available to measure rates of cell proliferation. These dyes have the advantage of providing greater sensitivity than colorimetric dyes, which can be a benefit when measuring smaller differences in cell proliferation rates between different cell populations.

Normal Cell Proliferation

Cells form the basic structural and functional units of an organism. All cells contain a cell membrane, cytoplasm and nucleus. Situated in the nucleus is the genetic material or deoxyribonucleic acid (DNA), which is the fundamental building block for life. DNA is made up of subunits called genes. Each gene is coded for a specific product such as a protein or enzyme.

Genes are contained in chromosomes and only the genes required are switched on. Some important genes in the context of cellular proliferation include:

- Proto-oncogenes: A gene involved in normal cell growth. Mutations (changes) in a proto-oncogene may cause it to become an oncogene, in which it becomes overactive and can cause the growth of cancer cells.

- Tumour suppressor genes: A type of gene that makes a protein called a tumour suppressor protein that helps control cell growth. Mutations (changes in DNA) in tumour suppressor genes may lead to cancer.

Each tissue and organ in the body is composed of vast populations of cells, totaling more than 10 (100 000 000 000 000). An astonishing 10 (1 000 000 000 000) cells die or are shed in the normal course of each day and must be replaced to sustain life.

The process by which cells grow and divide to replenish lost cells is termed cell proliferation. This is a highly regulated activity in normal, healthy tissue. The synthesis of new cells is balanced against cell loss so that the total number of cells composing all tissues and organs in the body remains essentially unchanged.

Cell growth, the replication of genetic material and cell division are all governed by the cell cycle, a highly-ordered series of events that culminates in mitosis (the division of a cell, giving rise to two daughter cells). Progression through the cell cycle depends on successful passage through a number of critical phases, known as checkpoints, which function to ensure the synthesis of fully functioning daughter cells.

Abnormal Cell Proliferation

While some cell types, such as those that compose the skin and bone marrow, continue to proliferate throughout life, other types including bone and muscle cells cease active proliferation when a human reaches adulthood. Most normal cells remain in a non-proliferative state unless they are stimulated to divide to replace lost cells. Abnormal regulation of the cell cycle can lead to the over proliferation of cells and an accumulation of abnormal cell numbers. Cancer cells arise from one cell that becomes damaged, and when divided, the damage is passed on to the daughter cell and again to the granddaughter cells and so on. Such uncontrolled, abnormal growth of cells is a defining characteristic of cancer.

The total number of cells composing the human body is determined not only by the rate of proliferation of cells but also by the rate of cell loss. Excess cells and those that are aged or have sustained damage that impairs normal functioning are eliminated to prevent accumulation of abnormal numbers of cells. The mechanism for regulating the removal of excess and impaired cells is known as apoptosis. Also referred to as cell suicide or programmed cell death, apoptosis is an orderly process during which internal cellular structures are progressively dismantled, the impaired cell shrinks and finally is rapidly destroyed by immune cells.

Role of Key Genes TP53 and RB1

A number of key genes, proteins and enzymes that regulate the cell cycle and the process of apoptosis have been identified. The TP53 gene and the regulatory protein it is responsible for producing, p53, together with the RB1 gene and its related protein, pRB, act to inhibit cell proliferation. In addition, the role of a group of proteins (cyclins and enzymes) known as cyclin-dependent kinases (CDKs) that act to stimulate a cell to progress through the cell cycle have also been identified.

It is now understood that mutations of these key genes affect the action of regulating proteins and enzymes and lead to the loss of regulation of cell proliferation that is seen in cancer. Mutations of the TP53 gene are also implicated in disturbances in apoptosis.

Cells with mutated TP53 genes evade the apoptosis mechanisms normally responsible for eliminating impaired cells.

DNA mutations may result from:

- Artificial sources (pesticides, organic chemicals, alkylating agents).

- Naturally occurring sources (plant toxins, viruses).

- Radiation.

When cell cycle control checkpoints fail, the following may occur:

- The mistake is quickly fixed.

- A mutation results in the production of an abnormal protein or enzyme.

- A mutation occurs near or around the proto-oncogene turning on cell division when not required.

- A mutation occurs near or around tumour suppressor gene (e.g. the p53 gene that normally inhibits the growth of tumours) resulting in inability to stop uncontrolled cell division.

Regulation of Cell Proliferation

Cell proliferation is regulated by the extracellular environment, such as the presence of nutrients, which plays a critical role in the proliferation of unicellular organisms. In multicellular organisms, however, nutrients are not the only extracellular environmental factor playing an important role. The internal environment of multicellular organisms is constantly and appropriately maintained, with energy and oxygen supplied, waste and CO_2 removed, and other factors such as temperature and pH maintained at appropriate levels. Despite such an optimal environment, cells in multicellular organisms do not proliferate uncontrollably, because cell proliferation is regulated in a more advanced manner.

Positive and Negative Control

It is known that tissues made of somatic cells include many cells that do not proliferate (i.e. in the Go phase) in normal circumstances, but proliferate only as needed. Cell proliferation in multicellular eukaryotes is regulated strictly through the balance of negative and positive cell proliferation signals. For instance, cells adhere closely to each other in the epithelial tissues of mammals, including humans, which controls proliferation negatively. If adhesion to the next cell is lost due to an injury such as a scratch on the skin, the negative control is lost—thus allowing proliferation. However, this event alone does not initiate proliferation. Cell proliferation is triggered only when the positive cell proliferation signals are sent to the cells by proliferative factors (or growth factors; many of which are proteins) that are specific to the cell type. If cells adhere properly to each other, proliferation is not initiated despite the presence proliferative factors. Thus, initiation of cell proliferation is regulated by positive and negative signals.

Signal transduction leading up to the initiation of cell proliferation.

Figure illustrates the intracellular signal transduction (signaling cascade) that starts when proliferation inducing signals (positive proliferation signals) reach the cell membrane receptor. Cell proliferation is initiated through this process. The transcription factor synthesized by a newly expressed early gene further induces the expression of various protein genes. Among the genes expressed as a result, here too, CDKs and cyclins are the most important. These two proteins are synthesized and accumulated to form cyclin-CDK complexes. If they are activated, they phosphorylate other proteins (such as the Rb protein), activate transcription factors (such as E2F) of the genes required for DNA synthesis, and finally activate enzymes that synthesize the DNA materials necessary in the S phase, thereby allowing the cell cycle transition to the S phase.

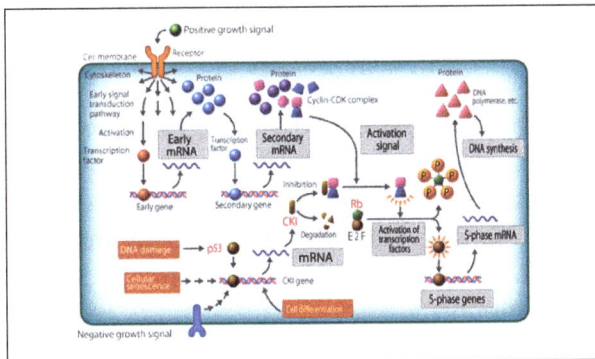

Signal transduction up to S-phase initiation.

Methods for Measuring Cell Proliferation

There are several methods available to measure cell proliferation rates. One method is to measure the overall metabolic activity inside a cell. Several dyes are available that can permeabilize a cell and react with certain enzymes and other factors and form a colored end-product which can be easily detected. One such dye, MTT, is a well-published molecule that produces a distinct purple color when added to proliferating cells. Another dye, WST-1, provides a similar color change in proliferating cells, but has the advantage of a single-step method that eliminates the detergent solubilization step required for MTT.

Fluorescence dyes are also available to measure rates of cell proliferation. These dyes have the advantage of providing greater sensitivity than colorimetric dyes, which can be a benefit when measuring smaller differences in cell proliferation rates between different cell populations.

Besides overall metabolic activity, cell proliferation may be measured by examining one or more specific markers within a cell.

Apoptosis

Apoptosis is a process that occurs in multicellular when a cell intentionally "decides" to die. This often occurs for the greater good of the whole organism, such as when the cell's DNA has become damaged and it may become cancerous.

Apoptosis is referred to as "programmed" cell death because it happens due to biochemical instructions in the cell's DNA; this is opposed to the process of "necrosis," when a cell dies due to outside trauma or deprivation.

Like many other complex cellular processes, apoptosis is triggered by signal molecules that tell the cell it's time to commit cellular "suicide."

The two major types of apoptosis pathways are "intrinsic pathways," where a cell receives a signal to destroy itself from one of its own genes or proteins due to detection of DNA damage; and "extrinsic pathways," where a cell receives a signal to start apoptosis from other cells in the organism. The extrinsic pathway may be triggered when the organism recognizes that a cell has outlived its usefulness or is no longer a good investment for the organism to support.

Apoptosis plays a role in causing and preventing some important medical processes. In humans, apoptosis plays a major role in preventing cancer by causing cells with damaged DNA to commit "suicide" before they can become cancerous. It also plays a role in the atrophy of muscles, where the body decides that it's no longer a good idea to spend calories on maintaining muscle cells if the cells are not being regularly used.

Because apoptosis can prevent cancer, and because problems with apoptosis can lead to some diseases, apoptosis has been studied intensely by scientists since the 1990s.

Function of Apoptosis

Apoptosis is an important evolutionary adaptation because it allows organisms to destroy their own cells. At first glance, that may sound like a terrible idea.

In the case of cells with damaged DNA that could become cancerous. Apoptosis is a major killer of pre-cancerous cells, and people with mutations that prevent apoptosis from functioning correctly are much more likely to get cancer. Multicellular organisms may also wish to lose cells that are no longer useful to the organism.

Examples of Apoptosis

From Tadpole to Frog

A spectacular example of this is found in frog tadpoles, which destroy and re-absorb entire body structures as they undergo their transformation into frogs.

Cells from tadpole's gills, fins, and tail are "told" to die by apoptosis signals as the tadpole matures. The raw materials of these dissembled cells become building materials and food for their new growing limbs.

Human Nervous System Development

During the early development of the human nervous system, huge numbers of cells die through apoptosis.

The truth is, scientists are not entirely sure why so much programmed cell death occurs in the developing nervous system. Some think it is because forming the correct connections is a complex and potentially difficult process for young neurons; and because maximum efficiency of the nervous system is definitely in the organism's best interest.

Nerves require huge amounts of energy in order to function, the nervous system consumes about 20-25% of all calories consumed in the human body.

Neurons also have to find their way to very precise targets. Early in development, neurons grow from furiously dividing stem cell "parents" and follow chemical signals to try to find the correct target cells to connect with. Connections must be formed between the brain and skin, between the brain and muscles, between neurons in the brain and rod and cone cells in the retina, etc.

To create this incredibly complex targeting, the developing nervous system simply grows way too many cells. Those that connect efficiently with the correct targets are used frequently, and they are preserved. But those that don't make contact efficiently and are not used frequently die off away by apoptosis.

Mouse Feet

During embryonic development, the feet of mice start out as flat, spade-shaped things. As development proceeds, the feet separate into five distinct toes by the process of apoptosis. Cells that connect the toes die off in order to create the distinct gaps between them.

This is an example of how programmed cell death can be used to shape useful structures and create useful features, in addition to getting rid of un-needed ones.

Apoptosis and Cancer

One primary function of apoptosis is to destroy cells that are dangerous to the rest of the organism. A common reason for apoptosis is when a cell recognizes that its DNA has been badly damaged. In these cases, the DNA damage triggers apoptosis pathways, ensuring that the cell cannot become a malignant cancer.

However, clearly this process sometimes fails. All instances of cancer are presumably instances where a damaged cell did not commit apoptosis, but instead went on to make more of itself.

Apoptosis may be unable to occur if essential genes required for it are among those that are damaged. However, some doctors and scientists have been studying apoptosis intensely in hopes that they may be able to learn to trigger it specifically in cancer cells using new medications or other therapies.

As with all drugs designed to kill cancer cells, the challenge with drugs designed to induce apoptosis is to ensure that these drugs only effect cancer cells. A medication that causes healthy cells as well as cancerous ones to commit programmed cell death could be very dangerous.

Apoptosis Pathway

There are two major types of apoptosis pathways, each of which illustrates an important point about how apoptosis is triggered and why it is useful.

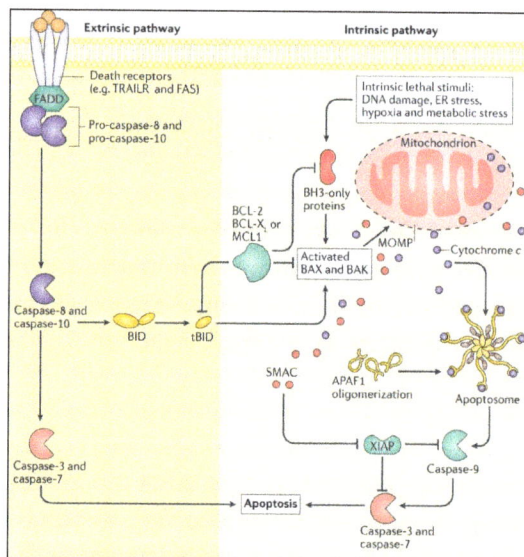

Extrinsic Pathway

In the "extrinsic" pathway to apoptosis, a signal is received from outside the cell instructing it to commit programmed cell death. This may occur if the cell is no longer needed, or if it is diseased.

Like many pathways for bringing about complex changes in a cell, the extrinsic pathway to apoptosis involves many steps, each of which can be "upregulated" or "downregulated" by gene expression or by other molecules:

Step 1

Like most signaling between cells, the extrinsic pathway of apoptosis starts with a signal molecule binding to a receptor on the outside of the cell membrane.

Two common types of chemical messengers that trigger the extrinsic pathway to apoptosis are FAS and TRAIL. These molecules may be excreted by neighboring cells if a cell is damaged or no longer needed.

The receptors that bind to FAS and TRAIL are called "FASR" for "FAS Receptor" or "TRAILR" for "TRAIL Receptor."

As with most receptor proteins, when FASR and TRAILR encounter to their signal molecule – sometimes called a "ligand" – they bind to it.

The binding process causes changes to the receptor's intracellular domain.

Step 2

In response to the changes in the intracellular domain of TRAILR or FASR, a protein inside the cell called FADD also changes.

FADD's name is either amusing or terrifying: it stands for "FAS-Associated Death Domain" protein.

Once FADD has been activated by changes to the receptor, it interacts with two additional proteins, which go on to start the process of cell death.

Step 3

Pro-caspase-8 and pro-caspase-10 are inactive proteins until they interact with an activated FADD. But if two of these molecules encounter an activated FADD, the parts of the proteins that keep them inactive are "cleaved" or "cut" away.

The pro-caspases then become caspase-8 and caspase-10 – which have been romantically referred to by scientists as "the beginning of the end" due to their role in starting apoptosis.

Caspases-8 and -10 disperse through the cytoplasm and trigger changes to several other molecules throughout the cell, including messengers that start the breakdown of DNA after being activated by the caspases.

Step 4

Another inactive molecule called BID is transformed into tBID when the activated caspases cleave off the part of BID that keeps the molecule inactive.

After BID is transformed into tBID, tBID moves to the mitochondria. tBID activates the molecules BAX and BAK.

The activation of BAX and BAK are the first steps shared by both the extrinsic and intrinsic pathways to apoptosis.

Steps 1-4 listed here are unique to the extrinsic pathway. But after BAX and BAK are activated, the subsequent steps are the same between both pathways.

As such, steps 3-7 of the intrinsic pathway, listed below, are also steps 5-9 of the extrinsic pathway.

Intrinsic Pathway

Step 1

The intrinsic pathway to apoptosis is triggered by stress or damage to the cell. Types of stress and damage that can lead the cell to apoptosis include damage to its DNA, oxygen deprivation, and other stresses that impair a cell's ability to function.

In response to these damages or stresses, the cell "decides" that its continued existence might be dangerous or costly to the organism as a whole. It then activates a set of proteins called "BH3-only proteins."

Step 2

BH3-only proteins are a class of proteins including several pro- and anti-apoptosis proteins. Apoptosis can be encouraged or discouraged, depending on which BH3-only proteins are activated or expressed.

Pro-apoptotic BH3-only proteins activate BAX and BAK – the same proteins that are activated by tBID after it is created through the extrinsic pathway to apoptosis.

Step 3

Activated BAX and BAK cause a condition known as "MOMP." MOMP stands for "mitochondrial outer membrane permeability."

MOMP is considered the "point of no return" for apoptosis. The steps leading up to MOMP can be stopped in their tracks by inhibitor molecules, but once MOMP has been achieved, the cell will complete the death process.

MOMP plays its key role in apoptosis by allowing the release of cytochrome C into the cytoplasm.

Step 4

Under normal circumstances, cytochrome C plays a key role in the mitochondrial electron transport chain. During MOMP, however, cytochrome C can escape the mitochondria and act as a signaling molecule in the cell cytoplasm.

Cytochrome-C in the cell cytoplasm prompts the formation of the ominous-sounding "apoptosome" – a complex of proteins that performs the final step to beginning cellular breakdown.

Step 5

The apoptosome, once it is formed, turns pro-caspase-9 into caspase-9.

Just as with the activation of caspases-8 and -10 in the extrinsic pathway to apoptosis, caspase-9 is able to trigger further changes throughout the cell.

Step 6

Caspase-9 performs several functions to promote apoptosis. Among the most important is the activation of caspases-3 and -7.

Step 7

Once activated, caspases-3 and -7 begin the breakdown of cellular materials. Caspase-3 condenses and breaks down the cell's DNA.

Time of Occurrence of Apoptosis

Apoptosis occurs when a cell's existence is no longer useful to the organism. This can occur for a few reasons.

If a cell has become badly stressed or damaged, it may commit apoptosis to prevent itself from becoming dangerous to the organism as a whole. Cells with DNA damage, for example, may become cancerous, so it is better for them to commit apoptosis before that can happen.

Other cellular stresses, such as oxygen deprivation, can also cause a cell to "decide" that it is dangerous or costly to the host. Cells that can't function properly may initiate apoptosis, just like cells that have experienced DNA damage.

In a third scenario, cells may commit apoptosis because the organism doesn't need them anymore due to its natural development.

One famous example is that of the tadpole, whose gill, fin, and tail cells commit apoptosis as the tadpole metamorphoses into a frog. These structures are needed when the tadpole lives in water – but become costly and harmful when it moves onto dry land.

Regulation of Apoptosis

Apoptosis occurs on a cell-by-cell basis. For each affected cell, two primary phases are observed: one of initiation and a second of execution. The resulting cell remnants are processed for reuse. Both phases are complex and require exquisite organization of multiple cellular systems, including interactions between proteins and cellular membranes. The initiation phase, or "death decision," became of significant interest following the description of a group of proteins in mammals known as the BCL-2 protein family. This protein family, which provides the framework for controlling apoptosis, takes its name from a type of cancer called B-cell lymphoma. BCL-2, the first family member, forms the molecular basis for sustaining the lymphoma cancer cells. The BCL-2 family of proteins has at least 25 members. Most of these are known as BH-3-only proteins. BH-3-only proteins function as activators or sensitizers of apoptosis and monitor important cell processes for dysfunction. They also control the function of two death-initiating, or pro-apoptotic, proteins (Bax and Bak) and a large number of death-preventing, or anti-apoptotic, proteins, which include BCL-XL and BCL-2. In mammals this control occurs primarily on the membranes of mitochondria, where the mortality decision for each cell is constantly reviewed under the supervision of the three contesting factions of the BCL-2 protein family.

A second family of proteins, the caspase proteolytic enzymes, contributes to both regulation by the BCL-2 family and execution of apoptosis after the death decision is confirmed. Caspases function in large part by the activation of other enzymes that dismantle the cellular cytoskeleton and cellular organelles and that degrade the nuclear DNA (deoxyribonucleic acid), from which there is no possibility of recovery. The cellular destruction occurs in place (in situ), and the cellular membrane system is reorganized to package the degraded components, including the digested genetic material, in membrane compartments. This packaging system prevents the materials from being released generally within tissues. Macrophages and other scavenger cells then engulf the membrane packets and process them for reuse as the most basic cellular components. In some instances, neighbouring cells may also engulf the degraded components.

Employing the intrinsic pathway, cancer cells, cells that are infected with bacteria or virus particles, and mutant cells can be assigned to apoptosis. The extrinsic pathway is commonly associated with cellular death receptors.

Cell-Cell Adhesion

Cell-cell signaling refers to inter-cellular communication through the transduction of chemical, mechanical or electrical signals, facilitated by the formation of specialized cell-cell adhesion junctions. There are three main types of cell-cell adhesive junctions in mammals that detect and transduce signals from neighbouring cells.

Tight Junctions

Also known as zonula occludens, tight junctions are found in epithelial and endothelial cells and primarily function as diffusion barriers. Integral membrane proteins commonly associated with tight junctions include; claudins, occludin, tricellulin and junctional adhesion molecules (JAMs). These proteins facilitate the formation of the anastomosing membrane strands that comprise tight junctions.

Adherens Junctions

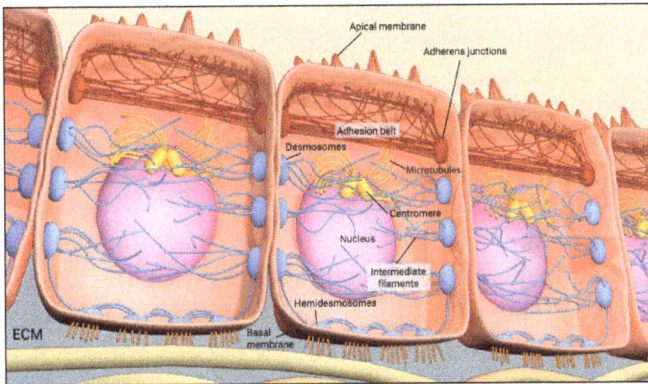

Adherens junctions link actin filaments between cells.

This diagram specifically depicts adherens junctions (red rectangles) connecting actin filaments (red lines) across polarized epithelial cells. In these cells, this results in the formation of contractile bundles of actin and myosin filaments near the apical surface – this forms a structure called the adhesion belt (arrows). Other junctions termed desmosomes (large blue rectangles) and hemi-desmosomes (small blue rectangles) link intermediate filaments (blue lines) between cells.

Adherens junctions regulate cell shape, maintain tissue integrity and translate acto-myosin-generated forces throughout a tissue. A key component of adherens junctions are the transmembrane glycoproteins cadherins, which bind the intracellular proteins p120-catenin and β-catenin, after which α-catenin is recruited by β-catenin. A vast network of adhesion receptors, scaffolding proteins, actin regulators and signaling proteins regulate the formation of this complex, via a complex network of interactions, which is currently being characterized and has been named the cadhesome.

The adherens junction is a type of stable cell contact, or anchoring junction, that keeps cells in a tissue together by forming an interconnected lateral bridge that links the actin cytoskeleton of neighboring cells. Adherens junctions come in many forms: In cardiac cells, contractile bundles are anchored by adherens junctions, while in nonepithelial tissues, adherens junctions link cortical actin filaments between cells.

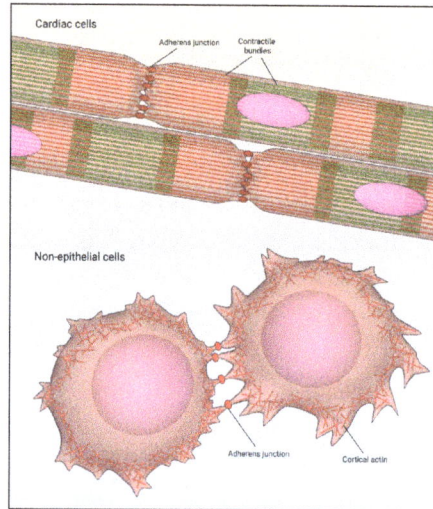

Adherens junction in cardiac cells and non-epithelial cells.

E-cadherin-mediated Cell-cell Adhesions

In polarized epithelial cells, the most obvious feature of the actin cytoskeleton is the contractile bundle of actin filaments usually located near the apical surface; this circumferential belt (aka adhesion belt) is linked from one cell to another through the extracellular domains of a continuous band of transmembrane molecules belonging to the cadherin family, along with other anchoring proteins (e.g. catenins, vinculin, and α-actinin).

However, recent investigation of cell-cell adhesion using super-resolution microscopy has uncovered new facts about the existence of the 'adhesion belt'. Conventional microscopy shows that clusters of E-cadherin merge to form a thick belt along the cell membrane between adjacent cells. The binding of individual E-cadherin proteins was thought to drive adhesion, with clusters formed in an adhesion- dependent manner, before merging and becoming uniformly distributed over time. This has long been the prevalent notion, based on conventional microscopy, which is limited in its ability to clearly visualize structures as small as the adherens junction or E-cadherin cluster.

Super-resolution imaging of the adherens junction revealed that cell-cell adhesions are initiated by small clusters of about five E-cadherin molecules. These distinct, evenly-sized, precursor E-cadherin clusters were observed in both incipient and mature cell-cell adhesions. As the adhesion develops, more E-cadherin molecules are recruited from neighbouring cells, leading to the clusters increasing in density. At the nanoscale level, it became clear that E-cadherin clusters never increased in size or merged to form the 'adhesion belt'. Instead, the actin cytoskeleton fences in E-cadherin clusters, preventing them from merging. The actin-delimited E-cadherin clusters therefore represent the building blocks of adherens junctions.

The actin and associated myosin are positioned to sense tension and to brace the cell Contraction of the actin network also provides motile forces for tissue morphogenesis and cell migration.

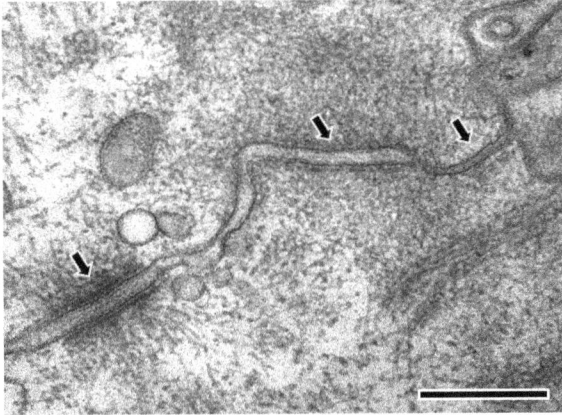

The cell is orientated such that the top right hand corner is apical and the bottom left hand corner is more basal. From left to right; the first arrow points to a desmosome with dense desmosomal plaques seen either side of the intercellular space of approximately 240 angstroms in width, the second arrow points to an adherens junction with an intercellular space of approximately 200 angstroms and the third arrow points towards a tight junction that shows close apposition of the plasma membrane leaflets, as inferred from the absence of an intercellular gap. (Scale bar 250 nm).

Desmosomes

These junctions link intermediate filaments (IFs) to the plasma membrane and in doing so provide mechanical stability to cells, which is particularly important for tissues and organs under high mechanical stress, such as the myocardium, skin, and bladder. Desmosomes comprise desmosomal cadherins (desmogleins and desmocollins) that facilitate linkage between apposing cells via their extracellular domains and bind other desmosomal components intracellularly via their cytoplasmic tails. Cytoplasmic desmosomal components include plakoglobin and plakophilins, which in turn bind intermediate filaments via desmoplakin.

In vertebrate epithelia, all three of the aforementioned junctions have defined spatial organizations, with tight junctions located most apically, followed by adherens junctions and lastly desmosomes. Each type of junction has a distinct ultrastructure, which can additionally vary in appearance dependent on cell type and morphology.

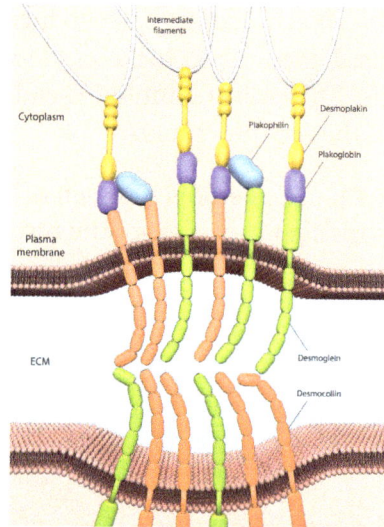

Desmosome junctions link intermediate filaments to the plasma membrane, and consist of desmosomal cadherins (desmogleins and desmocollins), and cytoplasmic desmosomal components such as plakoglobin and plakophilins, which connect the cadherins to the IF via desmoplakin.

Importance of Cadherins for Cell-Cell Adhesion

The calcium dependent cell adhesion molecules cadherins are usually 700-750 amino acids long. They are utilized in both desmosomes and adherens junctions to form protein complexes for cell-cell attachment. In desmosome complexes, a bridge is formed between the cadherins of each cell through homophilic binding.

This particular protein complex provides a strong adhesion useful for providing mechanical strength; therefore, desmosome complexes are abundant in heart and epidermis tissue that are frequently subjected to mechanical stress.

For adherens junctions, cadherins interact with actin filaments through intermediate anchor proteins called catenins. This type of cell-cell adhesion is common in epithelial and endothelial tissues.

Calcium has a pivotal role in cadherin functional activity by maintaining the rigid structure required for binding. In short, the calcium ions produce an inflexible structure through their positioning between each pair of cadherins, while the rigidity of the structure is positively correlated with the amount of calcium ions available.

Cell-cell Recognition and Cell-Cell Adhesion

Cell-cell recognition is an important part of the mechanism behind cell-cell adhesion. An early study noted the affinity found in cell adhesion molecules through the observation that cells from the same tissue preferentially adhere to one another. Cells disassociated from two separate organs can form a pellet when mixed, which gradually separate into cells derived from the same organ over time.

There are over one hundred different types of vertebrate cadherins and the experiment has been repeated with mixed cells sorted via cadherin type. It is these adhesive properties that allow for stable tissue architecture, albeit more recent assays have observed heterotypic binding affinity. It is believed that kinetic specificity allows cell adhesion molecules to recognize each other rather than broader thermodynamic specificity.

Selectins and Cell-Cell Adhesion

Selectins are another type of calcium-dependent cell adhesion molecules that allow cell-cell interactions within the blood stream. Different types of selectin are utilized for white blood cells, blood platelets and endothelial cells.

The cell-cell adhesions within the bloodstream are formed through the selectin binding to a specific oligosaccharide on the adjacent cell. The binding of white blood cells to the endothelial cells lining blood vessels is particularly important in allowing the transportation of cells from the blood stream into tissues.

This cell-cell interaction between white blood cells and endothelial cells is strengthened by integrins, another cell adhesion molecule. The combination of selectins and integrins provides the weak adhesion that allows white blood cells to roll along the surface of the blood vessel. However, when necessary, the two cell adhesion molecules can also bind the blood cell strongly to the endothelial cell surface for exit through the blood vessel between endothelial cells.

Immune System Biology

The immune system is designed to defend the body against foreign or dangerous invaders. Such invaders include:

- Microorganisms (commonly called germs, such as bacteria, viruses, and fungi).

- Parasites (such as worms).

- Cancer cells.

- Transplanted organs and tissues.

To defend the body against these invaders, the immune system must be able to distinguish between:

- What belongs in the body (self).

- What does not (nonself or foreign).

Antigens are any substances that the immune system can recognize and that can thus stimulate an immune response. If antigens are perceived as dangerous (for example, if they can cause disease), they can stimulate an immune response in the body. Antigens may be contained within or on bacteria, viruses, other microorganisms, parasites, or cancer cells. Antigens may also exist on their own—for example, as food molecules or pollen.

A normal immune response consists of the following:

- Recognizing a potentially harmful foreign antigen.

- Activating and mobilizing forces to defend against it.

- Attacking it.

- Controlling and ending the attack.

If the immune system malfunctions and mistakes self for nonself, it may attack the body's own tissues, causing an autoimmune disorder, such as rheumatoid arthritis, Hashimoto thyroiditis, or systemic lupus erythematosus (lupus).

Disorders of the immune system occur when:

- The body generates an immune response against itself (an autoimmune disorder).

- The body cannot generate appropriate immune responses against invading microorganisms (an immunodeficiency disorder).

- The body generates an excessive immune response to often harmless foreign antigens and damages normal tissues (anallergic reaction).

Components of the Immune System

The immune system has many components:

Antibodies (immunoglobulins) are proteins that are produced by white blood cells called B cells and that tightly bind to the antigen of an invader, tagging the invader for attack or directly neutralizing it. The body produces thousands of different antibodies. Each antibody is specific to a given antigen.

Antigens are any substance that the immune system can recognize and that can thus stimulate an immune response.

B cells (B lymphocytes) are white blood cells that produce antibodies specific to the antigen that stimulated their production.

Basophils are white blood cells that release histamine (a substance involved in allergic reactions) and that produce substances to attract other white blood cells (neutrophils and eosinophils) to a trouble spot.

Cells are the smallest unit of a living organism, composed of a nucleus and cytoplasm surrounded by a membrane.

Chemotaxis is the process of using a chemical substance to attract cells to a particular site.

The complement system consists of a group of proteins that are involved in a series of reactions (called the complement cascade) designed to defend the body—for example, by killing bacteria and other foreign cells, making foreign cells easier for macrophages to identify and ingest, and attracting macrophages and neutrophils to a trouble spot.

Cytokines are proteins that are secreted by immune and other cells and that act as the immune system's messengers to help regulate an immune response.

Dendritic cells are derived from white blood cells. They reside in tissues and help T cells recognize foreign antigens.

Eosinophils are white blood cells that kill bacteria and other foreign cells too big to ingest, and they may help immobilize and kill parasites and help destroy cancer cells. Eosinophils also participate in allergic reactions.

Helper T cells are white blood cells that help B cells produce antibodies against foreign antigens, help killer T cells become active, and stimulate macrophages, enabling them to ingest infected or abnormal cells more efficiently.

Histocompatibility (literally, compatibility of tissue) is determined by human leukocyte antigens (self-identification molecules). Histocompatibility is used to determine whether a transplanted tissue or organ will be accepted by the recipient.

Human leukocyte antigens (HLA) are a group of identification molecules located on the surface of all cells in a combination that is almost unique for each person, thereby enabling the body to distinguish self from nonself. This group of identification molecules is also called the major histocompatibility complex.

An immune complex is an antibody attached to an antigen.

An immune response is the reaction of the immune system to an antigen.

Immunoglobulin is another name for antibody.

Interleukin is a type of messenger (cytokine) secreted by some white blood cells to affect other white blood cells.

Killer (cytotoxic) T cells are T cells that attach to infected cells and cancer cells and kill them.

Leukocyte is another name for a white blood cell, such as a monocyte, a neutrophil, an eosinophil, a basophil, or a lymphocyte (a B cell or T cell).

Lymphocytes are the white blood cells responsible for acquired (specific) immunity, including producing antibodies (by B cells), distinguishing self from nonself (by T cells), and killing infected cells and cancer cells (by killer T cells).

Macrophages are large cells that develop from white blood cells called monocytes. They ingest bacteria and other foreign cells and help T cells identify microorganisms and other foreign substances. Macrophages are normally present in the lungs, skin, liver, and other tissues.

Major histocompatibility complex (MHC) is a synonym for human leukocyte antigens.

Mast cells are cells in tissues that release histamine and other substances involved in inflammatory and allergic reactions.

A molecule is a group of atoms chemically combined to form a unique substance.

Natural killer cells are a type of white blood cell that can recognize and kill abnormal cells, such as certain infected cells and cancer cells, without having to first learn that the cells are abnormal.

Neutrophils are white blood cells that ingest and kill bacteria and other foreign cells.

Phagocytes are a type of cell that ingests and kills or destroys invading microorganisms, other cells, and cell fragments. Phagocytes include neutrophils and macrophages.

Phagocytosis is the process of a cell engulfing and ingesting an invading microorganism, another cell, or a cell fragment.

A receptor is a molecule on a cell's surface or inside the cell that can identify specific molecules, which fit precisely in it—as a key fits in its lock.

Regulatory (suppressor) T cells are white blood cells that help end an immune response.

T cells (T lymphocytes) are white blood cells that are involved in acquired immunity. There are three types: helper, killer (cytotoxic), and regulatory.

Lines of Defense

The body has a series of defenses. It defenses include:

- Physical barriers,
- White blood cells,
- Molecules such as antibodies and complement proteins,
- Lymphoid organs.

Physical Barriers

The first line of defense against invaders is mechanical or physical barriers:

- The skin.

- The cornea of the eyes.

- Membranes lining the respiratory, digestive, urinary, and reproductive tracts.

As long as these barriers remain unbroken, many invaders cannot enter the body. If a barrier is broken—for example, if extensive burns damage the skin—the risk of infection is increased.

In addition, the barriers are defended by secretions containing enzymes that can destroy bacteria. Examples are sweat, tears in the eyes, mucus in the respiratory and digestive tracts, and secretions in the vagina.

Molecules

Innate immunity and acquired immunity interact, influencing each other directly or through molecules that attract or activate other cells of the immune system—as part of the mobilization step in defense. These molecules include:

- Cytokines (which are the messengers of the immune system).

- Antibodies.

- Complement proteins (which form the complement system).

These substances are not contained in cells but are dissolved in a body fluid, such as plasma (the liquid part of blood).

Some of these molecules, including some cytokines, promote inflammation.

Inflammation occurs because these molecules attract immune system cells to the affected tissue. To help get these cells to the tissue, the body sends more blood to the tissue. To carry more blood to the tissue, blood vessels expand and become more porous, allowing more fluids and cells to leave blood vessels and enter the tissue. Inflammation thus tends to cause redness, warmth, and swelling. The purpose of inflammation is to contain the infection so that it does not spread. Then other substances produced by the immune system help the inflammation resolve and damaged tissues heal. Although inflammation may be bothersome, it indicates that the immune system is doing its job. However, excessive or long-term (chronic) inflammation can be harmful.

Lymphoid Organs

The immune system includes several organs in addition to cells dispersed throughout the body. These organs are classified as primary or secondary lymphoid organs.

The primary lymphoid organs are the sites where white blood cells are produced and/ or multiply:

- The bone marrow produces all the different types of white blood cells, including neutrophils, eosinophils, basophils, monocytes, B cells, and the cells that develop into T cells (T cell precursors).

- In the thymus, T cells multiply and are trained to recognize foreign antigens and to ignore the body's own antigens. T cells are critical for acquired immunity.

When needed to defend the body, the white blood cells are mobilized, mainly from the bone marrow. They then move into the bloodstream and travel to wherever they are needed.

Lymphatic System: Helping Defend Against Infection

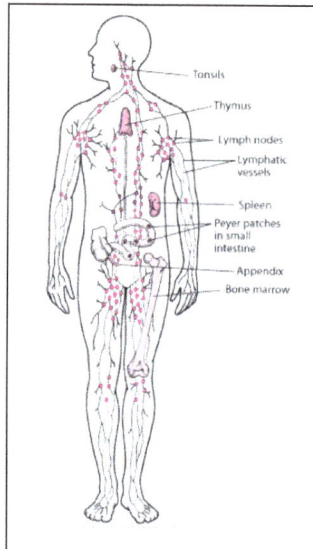

The lymphatic system is a vital part of the immune system, along with the thymus, bone marrow, spleen, tonsils, appendix, and Peyer patches in the small intestine.

The lymphatic system is a network of lymph nodes connected by lymphatic vessels. This system transports lymph throughout the body.

Lymph is formed from fluid that seeps through the thin walls of capillaries into the body's tissues. This fluid contains oxygen, proteins, and other nutrients that nourish the tissues. Some of this fluid reenters the capillaries and some of it enters the lymphatic vessels (becoming lymph).

Small lymphatic vessels connect to larger ones and eventually form the thoracic duct. The thoracic duct is the largest lymphatic vessel. It joins with the subclavian vein and thus returns lymph to the bloodstream.

Lymph also transports foreign substances (such as bacteria), cancer cells, and dead or damaged cells that may be present in tissues into the lymphatic vessels and to lymph nodes for disposal. Lymph contains many white blood cells.

All substances transported by the lymph pass through at least one lymph node, where foreign substances can be filtered out and destroyed before fluid is returned to the bloodstream. In the lymph nodes, white blood cells can collect, interact with each other and with antigens, and generate immune responses to foreign substances. Lymph nodes contain a mesh of tissue that is tightly packed with B cells, T cells, dendritic cells, and macrophages. Harmful microorganisms are filtered through the mesh, then identified and attacked by B cells and T cells.

Lymph nodes are often clustered in areas where the lymphatic vessels branch off, such as the neck, armpits, and groin.

The secondary lymphoid organs include the:

- Spleen,

- Lymph nodes,

- Tonsils,

- Appendix,

- Peyer patches in the small intestine.

These organs trap microorganisms and other foreign substances and provide a place for mature cells of the immune system to collect, interact with each other and with the foreign substances, and generate a specific immune response.

The lymph nodes are strategically placed in the body and are connected by an extensive network of lymphatic vessels—the lymphatic system. The lymphatic system transports microorganisms, other foreign substances, cancer cells, and dead or damaged cells from the tissues to the lymph nodes, where these substances and cells are filtered out and destroyed. Then the filtered lymph is returned to the bloodstream.

Lymph nodes are one of the first places that cancer cells can spread. Thus, doctors often evaluate lymph nodes to determine whether a cancer has spread. Cancer cells in a lymph node can cause the node to swell. Lymph nodes can also swell after an infection because immune responses to infections are generated in lymph nodes. Sometimes lymph nodes swell because bacteria that are carried to a lymph node are not killed and cause an infection in the lymph node (lymphadenitis).

Plan of Action

A successful immune response to invaders requires:

- Recognition,

- Activation and mobilization,

- Regulation,

- Resolution.

Recognition

To be able to destroy invaders, the immune system must first recognize them. That is, the immune system must be able to distinguish what is nonself (foreign) from what is self. The immune system can make this distinction because all cells have identification molecules (antigens) on their surface. Microorganisms are recognized because the identification molecules on their surface are foreign.

In people, the most important self-identification molecules are called:

- Human leukocyte antigens (HLA), or the major histocompatibility complex (MHC).

HLA molecules are called antigens because if transplanted, as in a kidney or skin graft, they can provoke an immune response in another person (normally, they do not provoke an immune response in the person who has them). Each person has an almost unique combination of HLAs. Each person's immune system normally recognizes this unique combination as self. A cell with molecules on its surface that are not identical to those on the body's own cells is identified as being foreign. The immune system then attacks that cell. Such a cell may be a cell from transplanted tissue or one of the body's cells that has been infected by an invading microorganism or altered by cancer. (HLA molecules are what doctors try to match when a person needs an organ transplant.)

Some white blood cells—B cells (B lymphocytes)—can recognize invaders directly. But others—T cells (T lymphocytes)—need help from cells called antigen-presenting cells:

- Antigen-presenting cells ingest an invader and break it into fragments.

- The antigen-presenting cell then combines antigen fragments from the invader with the cell's own HLA molecules.

- The combination of antigen fragments and HLA molecules is moved to the cell's surface.

- A T cell with a matching receptor on its surface can attach to part of the HLA molecule presenting the antigen fragment, as a key fits into a lock.

- The T cell is then activated and begins fighting the invaders that have that antigen.

T Cells Recognize Antigens

T cells are part of the immune surveillance system. They travel through the bloodstream and lymphatic system. When they reach the lymph nodes or another secondary lymphoid organ, they look for foreign substances (antigens) in the body. However, before they can fully recognize and respond to a foreign antigen, the antigen must be processed and presented to the T cell by another white blood cell, called an antigen-presenting cell. Antigen-presenting cells consist of dendritic cells (which are the most effective), macrophages, and B cells.

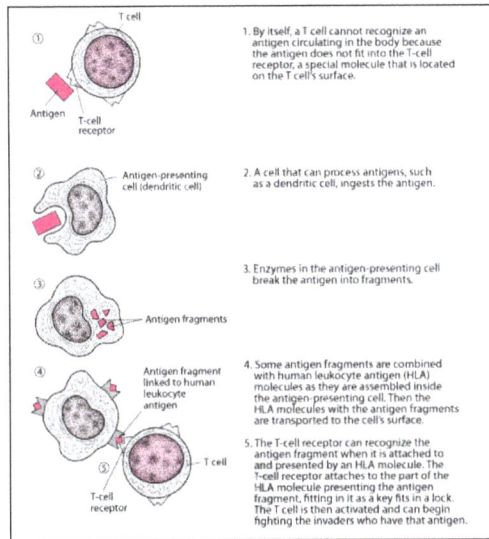

1. By itself, a T cell cannot recognize an antigen circulating in the body because the antigen does not fit into the T-cell receptor, a special molecule that is located on the T cell's surface.

2. A cell that can process antigens, such as a dendritic cell, ingests the antigen.

3. Enzymes in the antigen-presenting cell break the antigen into fragments.

4. Some antigen fragments are combined with human leukocyte antigen (HLA) molecules as they are assembled inside the antigen-presenting cell. Then the HLA molecules with the antigen fragments are transported to the cell's surface.

5. The T-cell receptor can recognize the antigen fragment when it is attached to and presented by an HLA molecule. The T-cell receptor attaches to the part of the HLA molecule presenting the antigen fragment, fitting in it as a key fits in a lock. The T cell is then activated and can begin fighting the invaders who have that antigen.

T cells are part of the immune surveillance system. They travel through the bloodstream and lymphatic system. When they reach the lymph nodes or another secondary lymphoid organ, they look for foreign substances (antigens) in the body. However, before they can fully recognize and respond to a foreign antigen, the antigen must be processed and presented to the T cell by another white blood cell, called an antigen-presenting cell. Antigen-presenting cells consist of dendritic cells (which are the most effective), macrophages, and B cells.

Activation and mobilization

White blood cells are activated when they recognize invaders. For example, when the antigen-presenting cell presents antigen fragments bound to HLA to a T cell, the T cell attaches to the fragments and is activated. B cells can be activated directly by invaders. Once activated, white blood cells ingest or kill the invader or do both. Usually, more than one type of white blood cell is needed to kill an invader.

Immune cells, such as macrophages and activated T cells, release substances that attract other immune cells to the trouble spot, thus mobilizing defenses. The invader itself may release substances that attract immune cells.

Regulation

The immune response must be regulated to prevent extensive damage to the body, as occurs in autoimmune disorders. Regulatory (suppressor) T cells help control the response by secreting cytokines (chemical messengers of the immune system) that inhibit immune responses. These cells prevent the immune response from continuing indefinitely.

Resolution

Resolution involves confining the invader and eliminating it from the body. After the invader is eliminated, most white blood cells self-destruct and are ingested. Those that are spared are called memory cells. The body retains memory cells, which are part of acquired immunity, to remember specific invaders and respond more vigorously to them at the next encounter.

Nonspecific Defense: The Innate Immune System

The human body has a series of nonspecific defenses that make up the innate immune system. These defenses are not directed against any one pathogen but instead, provide a guard against all infection.

First Line of Defense

The body's most important nonspecific defense is the skin, which acts as a physical barrier to keep pathogens out. Even openings in the skin (such as the mouth and eyes) are protected by saliva, mucus, and tears, which contain an enzyme that breaks down bacterial cell walls.

Second Line of Defense

If a pathogen does make it into the body, there are secondary nonspecific defenses that take place.

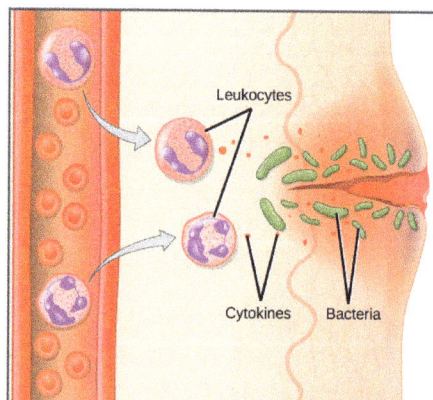

An inflammatory response begins when a pathogen stimulates an increase in blood flow to the infected area. Blood vessels in that area expand, and white blood cells leak from the vessels to invade the infected tissue. These white blood cells, called phagocytes engulf and destroy bacteria. The area often becomes red, swollen, and painful during an inflammatory response.

When a pathogen has invaded, the immune system may also release chemicals that increase body temperature, producing a fever. Increased body temperature may slow or stop pathogens from growing and helps speed up the immune response.

Specific Defense: The Adaptive Immune System

When pathogens are able to bypass innate immune defenses, the adaptive immune system is activated.

Cells that belong in the body carry specific markers that identify them as "self" and tell the immune system not to attack them.

Once the immune system recognizes a pathogen as "non-self," it uses cellular and chemical defenses to attack it. After an encounter with a new pathogen, the adaptive immune system often "remembers" the pathogen, allowing for a faster response if the pathogen ever attacks again.

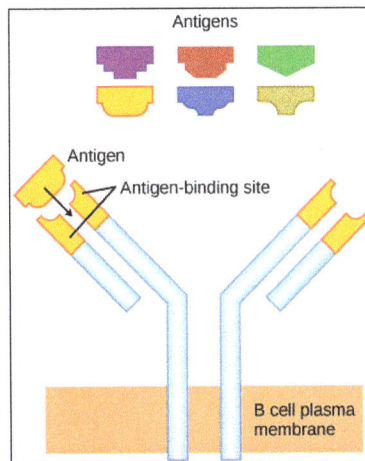

Specific immune responses are triggered by antigens. Antigens are usually found on the surface of pathogens and are unique to that particular pathogen. The immune system responds to antigens by producing cells that directly attack the pathogen, or by producing special proteins called antibodies. Antibodies attach to an antigen and attract cells that will engulf and destroy the pathogen.

The main cells of the immune system are lymphocytes known as B cells and T cells. B cells are produced and mature in bone marrow. T cells are also produced in bone marrow, but they mature in the thymus.

Humoral Immunity

Humoral immunity relies on the actions of antibodies circulating through the body.

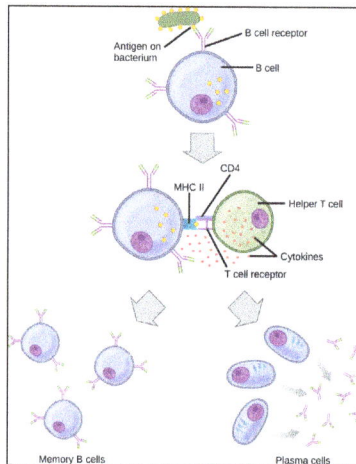

Humoral immunity begins when an antibody on a B cell binds to an antigen. The B cell then internalizes the antigen and presents it to a specialized helper T cell, which in turn activates the B cell.

Activated B cells grow rapidly, producing plasma cells, which release antibodies into the bloodstream, and memory B cells, which store information about the pathogen in order to provide future immunity.

Cell-mediated Immunity

Antibodies alone are often not enough to protect the body against pathogens. In these instances, the immune system uses cell-mediated immunity to destroy infected body cells.

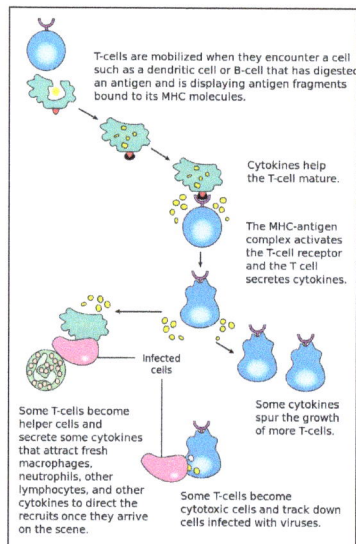

T cells are responsible for cell-mediated immunity. Killer T cells (cytotoxic T cells) assist with the elimination of infected body cells by releasing toxins into them and promoting apoptosis. Helper T cells act to activate other immune cells.

Vaccines

Vaccines work by taking advantage of antigen recognition and the antibody response. A vaccine contains the antigens of a pathogen that causes disease. For example, the smallpox vaccine contains the antigens specific to smallpox. When a person is vaccinated against smallpox, the immune system responds by stimulating antibody-producing cells that are capable of making smallpox antibodies. As a result, if the body comes into contact with smallpox in the future, the body is prepared to fight it.

Viral Structure

Viruses are infectious particles that reproduce by hijacking a host cell and using its machinery to make more viruses.

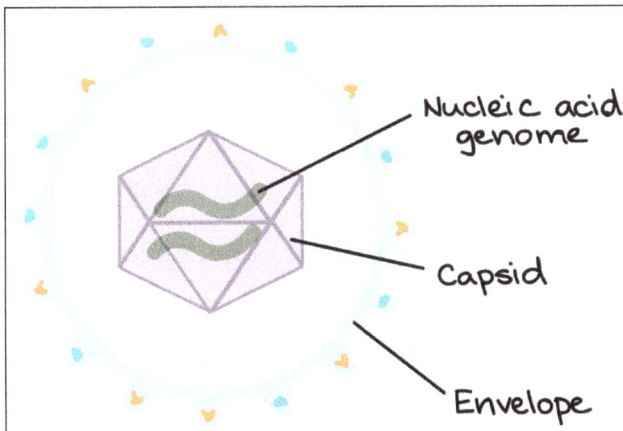

There are many kinds of viruses, differing in structure, genome, and host specificity. However, viruses tend to have several features in common. All viruses contain a protective protein shell, or capsid, that houses their nucleic acid genome (either DNA or RNA).

Some viruses also have a membrane layer called an envelope that surrounds the capsid.

Steps of Viral Infection

Viruses reproduce by infecting their host cells, providing instructions in the form of viral DNA or RNA, and then using the host cell's resources to make more viruses.

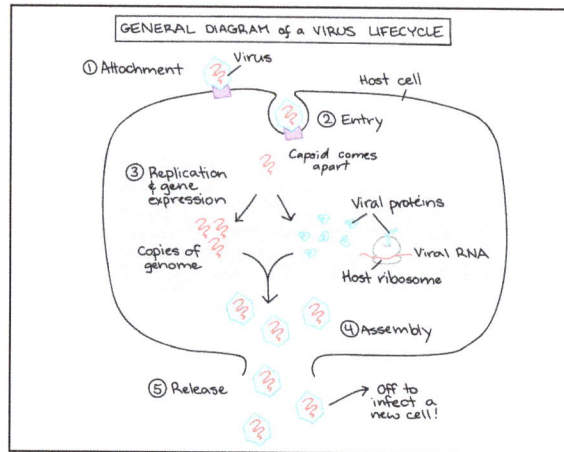

1. The virus recognizes and binds to a host cell via a receptor molecule on the cell surface.

2. The virus or its genetic material enters the cell.

3. The viral genome is copied and its genes are expressed to make viral proteins.

4. New viral particles are assembled from the genome copies and viral proteins.

5. Completed viral particles exit the cell and can infect other cells.

References

- Methods-measuring-cell-proliferation: cellbiolabs.com, Retrieved 8 July, 2019

- Normal-cell-proliferation, biology-of-cancer: edcan.org.au, Retrieved 6 February, 2019

- Abnormal-cell-proliferation, biology-of-cancer: edcan.org.au, Retrieved 13 April, 2019

- Methods-measuring-cell-proliferation: cellbiolabs.com, Retrieved 3 August, 2019

- Apoptosis: biologydictionary.net, Retrieved 19 June, 2019

- Apoptosis, science: britannica.com, Retrieved 29 March, 2019

- What-are-cell-cell-adhesions, what-is-mechanosignaling: mechanobio.info, Retrieved 22 January, 2019

- What-is-Cell-Cell-Adhesion, life-sciences: news-medical.net, Retrieved 11 July, 2019

- Immune-system, immune-disorders: msdmanuals.com, Retrieved 21 January, 2019

- Hs-the-immune-system, human-body-systems, biology, science: khanacademy.org, Retrieved 7 May, 2019

DNA Technologies

There are many technologies which closely work with DNA such as polymerase chain reaction and DNA cloning. Polymerase chain reaction (PCR) is a method within the field of molecular biology where many copies of a particular DNA segment are made. The diverse DNA technologies such as nucleic acid hybridization and DNA repair have been thoroughly discussed in this chapter.

DNA Cloning

Deoxyribonucleic acid (DNA) cloning is a cornerstone of molecular biology. Its power is such that without it a large proportion of modern biological science would simply not be possible.

A clone can be defined as an identical copy. DNA cloning involves joining, or 'ligating', DNA with a 'vector' that enables the resulting construct to be introduced into a cell, replicated and passed on to daughter cells as that cell divides. The result is a population of cells all containing clones of the original DNA. Thereafter that DNA is immortalized, since cells can be grown to order and DNA extracted for study or further manipulation whenever it is required.

DNA cloning has been made possible by the discovery of two types of protein, those that break or modify DNA such that they have suitable termini for ligation and those that are capable of ligating those molecules of DNA.

Methods of Cloning

DNA exists as a double helix of antiparallel polymer strands with the nucleotide units in each joined by 50 – 30 phosphodiester bonds. The ability to selectively break and join these bonds is the basis of DNA cloning.

Ligation

DNA strands can be joined through the action of the enzyme DNA ligase, the normal biological role of which is to join the series of Okazaki fragments that are generated on the lagging strand during replication. Typically the ligase from the Escherichia coli bacteriophage T4, known as 'T4 DNA ligase', is used in cloning experiments as it is both easily purified and highly active. Ligase can be purchased commercially and is normally supplied with a

10 × concentrate of reaction buffer, containing Mg^{2+} and adenosine 5' triphosphate (ATP) that are both required for the activity of this enzyme. Ligase utilizes the Mg^{2+} cation to contact the free 3' hydroxyl group at the end of a DNA strand and uses the energy stored in the phosphodiester bonds of ATP to covalently attach this position to the 5' phosphate at the terminus of another DNA strand, so joining the two strands. Ligating double-stranded DNA thus consumes two molecules of ATP and requires the two DNA molecules to have compatible termini. However, only one of the two 5' termini needs to have a phosphate group. This is useful because dephosphorylation of the vector DNA prevents its self-ligation. DNA can be ligated with either blunt ends or, more efficiently, as compatible staggered ('sticky'), cohesive ends, the conditions used being slightly different in each case.

For sticky-end ligation, insert DNA is incubated in 16 ligase buffer in the presence of approximately equimolar amounts (typically 10–50 ng) of vector that has compatible termini, with 0.1–1 units of T4 DNA ligase in a total volume of 10–50 µl for 2–8 h at 16 °C, or overnight at 4 °C. DNA molecules possessing sticky ends transiently base pair while in solution so bringing their ligatable termini into close proximity. It is for this reason that sticky-end ligation is more efficient.

Blunt-end ligation can be driven by increasing the concentration of termini such that ligation is favoured. Typically vector is mixed with a threefold molar excess of insert DNA in 1× ligase buffer, with 1 unit of T4 DNA ligase, in a total volume of 4–10 µl and incubated overnight at 16 °C.

Preparation of DNA for Ligation

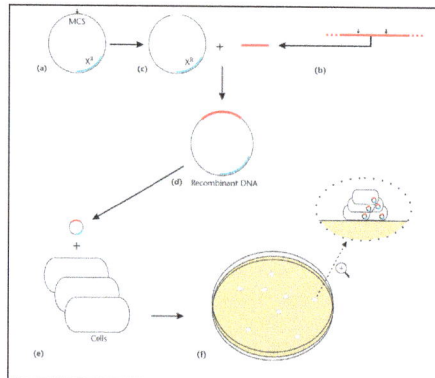

A schematic representation of a typical cloning experiment. (a) The vector is cut (;) within its multicloning site (MCS). (b) The target DNA is cut (;) so as to produce termini compatible with the vector. (c) The vector and insert are ligated to produce recombinant DNA. (d, e) Recombinant DNA is introduced into appropriate host cells. In this illustration the vector encodes resistance to an antibiotic, X. (f) If the cells are plated out onto medium containing X, only cells that have been transformed will grow and divide to form colonies (groups of around a million cells or 'clones' that have arisen from the same original transformed cell).

Prior to ligation, compatible insert and vector DNA must be prepared. Both need compatible termini and must be free of contaminating chemicals such as phenol, ethylenediaminetetraacetic acid (EDTA) or ethanol that may inhibit ligation. Vector DNA is typically prepared by digestion with the relevant restriction enzyme, then dephosphorylated by incubation with alkaline phosphatase to remove the terminal 5′ phosphate groups and so prevent vector self-ligation. Insert DNA is typically prepared using one of the methods described below, then if it is part of a mixture of DNA fragments, it is purified by separating that mixture on a preparative (normally low-melting-point) agarose gel, cutting out the slice of agarose containing the fragment one wishes to clone and extracting the DNA from the agarose matrix.

Extraction of DNA from agarose can be performed by one of a number of methods including melting followed by phenol extraction, agarase digestion of the agarose backbone or absorption and subsequent elution of the DNA from a silica matrix.

Generation of Sticky Ends

DNA molecules with sticky ends are usually generated through the use of type II restriction endonucleases. These enzymes, whose biological function is to digest or 'restrict' foreign DNA that enters the cell, recognize specific sequences, 'recognition sites' and catalyze the double-stranded cleavage of DNA within that site to generate either molecules with protruding 5′, protruding 3′ or blunt termini.

Some restriction enzymes generate protruding termini that are compatible, for example BamHI, BclI, BglII, DpnII, MboI and Sau3AI all cleave to give GATC 5′ overhangs, and so DNA cleaved by any of these enzymes will ligate to that cleaved by any of the others. This is of particular use when one wishes to generate genomic libraries. A DNA library is a ligation that contains a mixture of DNA inserts. For wholegenome studies, libraries are often made so as to contain inserts representing all sequences in that genome. The patterns of cleavage of a restriction enzyme with a six-base (and longer) recognition sequence are often nonrandom across a whole genome. Therefore a common technique is to digest the DNA with an enzyme with a four-base recognition site such as Sau3AI (GATC). Normally such an enzyme would cut every 4^4 (that is, 256), bases, but if the digestion time is kept short or the amount of enzyme limited, then a partial digestion will occur and so by controlling the conditions a spectrum of fragments in the desired size range can be produced. Since it is important to cleave the vector DNA only once, preferably within the desired cloning site region, and Sau3AI would digest a typical vector molecule into many pieces, the vector can be prepared using an enzyme such as BamH I that cuts less frequently but leaves compatible sticky ends.

Generation of Blunt Ends

DNA with blunt ends can be prepared using restriction enzymes that leave a blunt end, through cleavage with DNaseI or after end-repair of DNA molecules that have staggered termini. Although less efficient, this method has more versatility as there is no absolute requirement for the presence of a particular recognition site. DNA can be broken randomly by mechanical breakage or digested with any enzyme, then repaired. For ends with protruding 5′ termini, repair is commonly achieved by incubation with nucleotides and the Klenow fragment of E. coli DNA polymerase I. Protruding 3′ ends are removed by taking advantage of the high 3′ −5′ exonuclease activity of T4 DNA polymerase which digests the single-stranded bases in the overhang. If the nature of the ends is unknown, as would be the case for random breakage, then ends are normally repaired by digesting protruding singlestranded termini by incubation with mung bean or Bal3I nuclease or by incubating with a mixture of both T4 and Klenow polymerase.

Table: A selection of restriction endonucleases commonly used in DNA cloning.

Enzyme[a]	Source	Recognition site[b]	Ends produced
BamHI	Bacillus amyloliquefaciens H	5′ G^GATCC 3′	5′ overhang
EcoRI	Escherichia coli	5′ G^AATTC 3′	5′ overhang
HindIII	Haemophilus influenzae	5′ A^AGCTT 3′	5′ overhang
NcoI	Nocardia corallina	5′ C^CATGG 3′	5′ overhang
NdeI	Neisseria denitrificans	5′ CA^TATG 3′	5′ overhang
NotI	Nocardia otitidis-caviarum	5′ GC^GGCCGC 3′	5′ overhang
PstI	Providencia stuartii	5′ CTGCA^G 3′	3′ overhang
Sau3AI	Staphylococcus aureus 3A	5′ ^GATC 3′	5′ overhang
SmaI	Serratia marcescens	5′ CCC^GGG 3′	Blunt

Enzymes commonly get their names from abbreviations of the organism from which they were first isolated. Roman numerals after the name depict the order in which enzymes were discovered from an organism, for example, HindIII was the third enzyme to be isolated from Haemophilus influenzae.

The recognition sequence for one strand in the 5′ −3′ direction is shown. Typically restriction sites have twofold dyad symmetry, so the same site is seen when reading in the 5′ −3′ direction on the other strand. Enzymes cleave the DNA strand within each recognition site to give a 3′ hydroxyl group and a 5′ phosphate group, at the position denoted by.

Conversion of Blunt to Sticky Ends

There are several methods by which blunt ends can be given sticky tails:

- Through the action of the enzyme terminal transferase, tracts of a single nucleotide base can be added to the 3′ hydroxyl group at each terminus. If different but complementary bases are added to both vector and insert, then the two will be compatible and can be ligated.

- Single base overhangs can be added to the 3′ position of blunt termini by the action of Taq DNA polymerase. Taq DNA polymerase, commonly used in the polymerase chain reaction (PCR), frequently adds an extra adenosine at the 3' position of the amplified DNA. Thus PCR products can readily be cloned by ligating to a thymine tailed vector, socalled TA cloning. By incubating DNA with blunt termini in the presence of Taq DNA polymerase and a single deoxynucleotide triphosphate at 72 °C, DNA with single base 3' overhangs can easily be prepared.

- Short double-stranded adaptor or linker DNAs containing restriction enzyme recognition sequences can be ligated to blunt termini. Adaptors have one blunt end and one sticky end; thus when ligated they convert the blunt ends to sticky ones. Linkers have two blunt ends but contain internal restriction sites. After linker addition, the resulting molecules can be digested with the appropriate restriction enzyme and sticky ends created.

Generation of deoxyribonucleic acid for cloning by polymerase chain reaction.

PCR is an extremely powerful technique as it allows for minute quantities of DNA (or messenger ribonucleic acid (mRNA)) to be amplified up to cloneable quantities. PCR products synthesized using enzymes such as Taq DNA polymerase (that lack 3' −5' exonuclease activity) have a single adenosine 3' overhang and those synthesized with 'proofreading' polymerases which possess 3' −5' exonuclease activity are generated with blunt ends. Furthermore, since the primers used for PCR are incorporated into the products (and mismatches between primer and target sequence are often tolerated, especially at the 5' end of the primer) PCR can be used to introduce restriction sites for cloning at the termini of the amplified fragments. Of particular use in this respect are the recognition sites for the enzymes NdeI and NcoI, which contain the initiation codon sequence ATG. Here PCR can be used to introduce an NdeI or NcoI site at the initiation codon of a particular gene so that it can be cloned in-frame into an expression vector.

Vectors

Vectors are DNA molecules that are capable of accepting selected fragments of DNA and replicating the resulting hybrid when it is introduced into living cells. Typically vectors are modified versions of naturally occurring independent replicons or viruses. Due to its extremely well-characterized genetic systems, the majority of vectors propagate in laboratory strains of the enteric bacterium E. coli, although 'shuttle vectors' that can also replicate in other organisms and vectors that exclusively replicate in other organisms are widely available. Vectors can be obtained commercially from molecular biology reagent suppliers, from authors of published papers or from collections such as ATCC.

Features of Vectors

Cloning Site

The vector must possess a recognition site for at least one restriction endonuclease so that the vector can accept DNA. Most common vectors have been constructed so as to have a multicloning site (MCS). Here a short piece of DNA containing the recognition sequences for several enzymes has been added into the region of the vector where foreign DNA is to be inserted.

Origin of Replication

To be able to replicate and propagate the DNA construct, vectors must contain origin of replication (ori) sequences that are recognized by the host cell DNA replication systems. It is the ori that largely determines vector copy number.

Selectable Marker

A vector must contain sequences that confer a positive advantage on cells that contain it. This is because maintenance and propagation of cloned DNA places a burden upon that cell, and without an advantage cells that did not contain the clone would grow preferentially resulting eventually in loss of the clone from the population. Furthermore, when DNA is introduced into a cell population only a small fraction of those cells take up the DNA and one needs a mechanism to select for those cells that contain the vector construct. Therefore vectors are designed with selectable markers. These typically fall into two main classes:

- Genes that encode resistance to a certain antibiotic (e.g. ampicillin, G418, kanamycin, chloramphenicol, tetracycline, etc.).

- Auxotrophic markers which complement a host cell mutation in an essential biosynthetic pathway. For example, the yeast URA3 gene is commonly used in yeast vectors in combination with a host cell that has a mutation resulting in a requirement for uracil. When cells are grown in a medium that does not contain uracil only those containing the vector with the URA3 marker can grow.

Optional Features

There are a great many vectors now available to the molecular biologist. Each has slightly different features allowing it to be more suitable for one application than another, and so a careful look at the features that a vector contains is essential. Common features are noted below.

Sequences that Enable Recombinant Selection

Many vectors have their cloning sites within the coding region of a gene that confers a selectable phenotype, such that when DNA is inserted the function of that gene is disrupted. Since the vector can self-ligate to give nonrecombinant molecules, this is a valuable means of selecting for cells that contain recombinant clones. The most widely available selection is a-complementation, where the cloning site is located within coding sequences for the a fragment of the enzyme b-galactosidase (encoded by the lacZ gene). The a fragment is able to complement a host cell mutation (lacZDM15) in this gene. Thereby on selective agar plates containing the colorimetric substrate for this enzyme (X-gal) and the inducer isopropyl-b-D-thiogalactopyranoside (IPTG), nonrecombinants have b-galactosidase activity, break down X-gal into a blue compound and appear blue whereas recombinants where this gene has been disrupted remain white/colorless. Another recombinant selection method utilizes ccdB (or the cell-death gene) the product of which strongly inhibits bacterial DNA gyrase and thereby DNA synthesis, resulting in cell death. If the cloning site is positioned within this gene, only cells containing vector with an insert will grow. Likewise the sacB gene (present in P1 and some bacterial artificial chromosome (BAC) vectors) encodes the enzyme levansucrase, which can convert the common sugar sucrose into a cytotoxic compound. When cells are grown on medium containing sucrose, only recombinants where DNA has been inserted into the cloning site to disrupt this gene will grow.

Primer Binding Sites

Sequences that bind well-characterized (often commercially available) primers are often found flanking the multicloning site. These are useful for sequencing, or amplification by PCR, of the inserted DNA.

Reporter Genes

Reporter genes encode a product with an easily detectable activity such as lacZ (b-galactosidase), chloramphenicol acetyltransferase (CAT) or green fluorescent protein (GFP). The promoter activity of a given piece of DNA can be quantified by cloning that sequence upstream of a promoterless reporter gene. Likewise if a gene being studied is cloned in-frame with a reporter gene, the activity of the reporter gene will give a measure of the expression level of the corresponding gene product. In the case of GFP, information regarding the subcellular localization of the product can also be obtained.

Strong Promoter Sequences

1. Riboprobe generation. Strong RNA polymerase promoters (e.g. SP6, T3 or T7) often flank cloning sites; this enables single-stranded probes from the termini of the inserted DNA to be generated. This can be carried out in vitro when the construct is incubated with the corresponding polymerase and ribonucleotides.

2. Overexpression. For overexpression of a gene, for example, as a prelude to protein purification, that gene can be cloned into a vector that has a known strong promoter sequence, such as T7 polymerase promoter, Ptac, ParaBAD or PCMV, just upstream of the cloning site. These promoters are often inducible such that expression can be induced at the right time. In some instances the overexpressed protein can accumulate to over 50% of the total cell protein.

In-frame Genes Encoding Easily Purifiable Proteins

To facilitate purification of an overexpressed protein, the corresponding gene can be cloned in-frame with sequences that code for an easily purifiable protein or peptide such as glutathione-S-transferase (GST), maltose binding protein (MalE) or polyhistidine peptide (His-TAG). This results in the overexpression of a fusion protein comprising the protein under study and the easy to purify protein, which can be at either the N or C-terminus depending on vector and cloning site used. Subsequently this fusion can be readily purified by virtue of the characteristics of the particular fusion partner. His-TAG, which is probably the most commonly used fusion, readily binds nickel ions, and so His-TAGged fusions can be purified by affinity chromatography on immobilized nickel, washed, then subsequently eluted with imidazole. The fusion partner often has no effect on the activity of the protein and so is not normally removed, though some vectors incorporate protease cleavage sites such as that for factor Xa, at the fusion junction.

loxP Sites

loxP is the recognition sequence for the bacteriophage P1 recombinase enzyme, Cre, which catalyzes recombination between sequences that are flanked by such sites. By cloning DNA into vectors that possess loxP sites that DNA can, for example, be recombined into another vector containing loxP sites or into the host chromosome, when in the presence of either the Cre recombinase in vitro or a plasmid carrying the cre gene in vivo.

Cos Sites

cos sites are the recognition sequences for phage packaging systems. The DNA present between two such sites, provided it is within the packaging limits for that system, will be packaged into phage particles.

Types of Vectors

An outline of the features of each vector type is given in table.

Table: A summary of the properties of the various classes of vectors used in DNA cloning.

Vector	Cloning capacity	Use	Common examples	Method of introduction into cell	Type of growth on agar plate	Copy number
Plasmids	Up to 12 kb	Multiple (e.g. mutagenesis, overexpression, sequencing, screening, manipulation)	pUC, pBR322, pGEM	Transformation or electroporation into bacteria. Various methods for eukaryotes (see text)	Colonies	1–700

M13	Normally up to 6 kb	Probe generation, sequencing, in vitro mutagenesis	M13mp18	Transfection	Plaques	100–200
Bacteriophage lambda (l)	9–23 kb	Genomic and cDNA libaries, expression screening	lZAP, lFIX	Packaging of construct followed by infection	Plaques	N/A
Cosmids	Up to 47 kb	Genomic libraries, genome mapping	Supercos, lawrist, tropist	Packaging of construct followed by infection	Colonies	Up to 50
Fosmids	Up to 43 kb	Genomic libraries, genome mapping	pFOS1	Packaging of construct followed by infection	Colonies	1–2
P1 bacteriophage	30–100 kb	Genomic libraries, genome mapping	pAd10sac-BII	Packaging of construct followed by infection	Colonies	1–2
BACs	Up to 300 kb	Genomic libraries, genome mapping	pBACe3.6, pBeloBA-CII	Electroporation	Colonies	1–2
PACs	Up to 300 kb	Genomic libraries, genome mapping	pCYPAC2	Electroporation	Colonies	1–2
YACs	20–2000 kb	Genomic libraries, genome mapping	pYAC4	Electroporation, spheroplast absorption, or lithium-mediated transformation	Colonies	1

Plasmids

Plasmids are the most widely used vectors. This is probably because of the ease with which plasmid DNA can be isolated from the host cell and the variety of plasmids that are available. They are double-stranded independent replicons that range in size from approximately 2 to 10 kb and which are capable of accepting DNA inserts up to about 12 kb, though up to 5 kb is perhaps more common. Figure shows the structure of pUC18, a typical plasmid that replicates in E. coli.

M13 Bacteriophage

M13 is a filamentous bacteriophage, with a 6.4 kb single-stranded genome, that infects E. coli. An infected cell produces 100–200 double-stranded (or replicative form (RF)) copies of the phage genome. These encode genes that coordinate packaging of one of the phage genome DNA strands into a proteinaceous viral coat and their extrusion from the cell where they are then free to infect further cells. A series of M13-based cloning vectors have been developed, and found capable of accepting inserts several times larger than the M13 genome. However, DNA in M13 can be unstable and so insert sizes are normally kept below that of the vector and this system is not used for the longer-term maintenance of DNA. M13 has no selectable marker; instead infected cells are identified by their ability to form zones of clearing, or 'plaques'. If bacterial cells are mixed with a soft agar and allowed to grow on top of a normal agar plate, the cells in the soft agar will grow and give the agar layer an opaque appearance. Infected cells grow a lot more slowly than noninfected ones and so can be visualized as a plaque in the bacterial lawn.

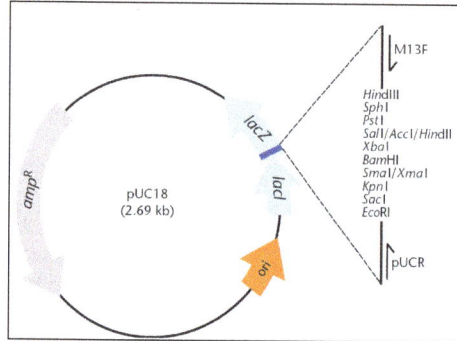

Figure: Diagram of the plasmid cloning vector, pUC18.

This 2.69 kb double-stranded, closed-circular DNA molecule has an origin of replication (ori) allowing multicopy propagation in E. coli, a b-lactamase gene (ampR) giving resistance to the antibiotic ampicillin and a lacI gene, the product of which induces the expression of the lacZ gene, which in turn encodes b-galactosidase. The multicloning site (dark bar, shown expanded to the right) is close to the 50 end of the lacZ gene, such that insertion at any of the points within the multicloning site (MCS) should disrupt b-galactosidase expression, thereby allowing blue/white recombinant selection on media containing the colorimetric substrate X-gal and inducer isopropyl-b-D-thiogalactopyranoside (IPTG). The MCS contains closely nested restriction sites for a number of enzymes, in the order shown. It is flanked by sequences to which the M13 forward and pUC reverse primers bind, facilitating polymerase chain reaction amplification or sequencing of the inserted DNA.

Since the extruded phage contain copies of one particular DNA strand, this system has proved very useful for techniques that utilize a single-stranded template such as sequencing, site-directed mutagenesis and producing radiolabeled probe DNA for hybridization.

Bacteriophage Lambda

Lambda (λ) is a naturally occurring phage virus of E. coli. It has a 48 kb linear double-stranded genome, with 12-base cohesive ends (cos sites) at each 50 end, inside an icosahedral proteinaceous head to which is attached a long noncontractile tail. Infection occurs as a result of the tail binding to the outer membrane maltose receptor and injection of the phage genome through the tail and into the cell. Once in the cell the DNA can follow a lytic or lysogenic pathway depending on environmental or intracellular conditions and the genotype of the individual phage. In the lysogenic pathway, the phage DNA is integrated into the E. coli genome providing the geneticist with a tool to integrate DNA into genomic DNA. In the lytic pathway, the DNA is replicated under a rolling circle mechanism to produce a long concatemer of around 100 lambda genome units. Lambda genes encode head and tail proteins and an enzyme that recognizes the cos sites and cleaves the concatemer into individual genomes. Subsequently each 48 kb unit is packaged into lambda heads, the tails attach and the infected cell lyses to release approximately 100 infective phage. In order to be packaged the cos sites need to be around 38– 52 kb apart and so different lambda vectors have been produced where nonessential regions have been deleted to make space for DNA to be inserted. A typical lambda vector λZAPII. There are two main advantages to using lambda vectors:

- The packaging and infection process is very efficient (perhaps 100 times more efficient than transformation/transfection).

- Since the cells die during lysis, lambda can maintain some inserts that carry cytotoxic or unstable/poorly tolerated sequences.

Cosmids/Fosmids

Cosmids were created by engineering cos sites into plasmid vectors to create hybrid vectors with the highefficiencypackaging/infectionadvantagesof thelambda system and the ease of isolation advantage of plasmids. In recent years, cosmids have been superseded as the vector of choice for creating genomic DNA libraries by artificial chromosome vectors that can accommodate larger inserts. Since they are maintained at quite a high copy number in the cell, cosmid clones can become unstable and are often prone to deletion. To overcome this, fosmids, which are based on the singlecopy bacterial F factor, have been developed.

P1 Bacteriophage Vectors

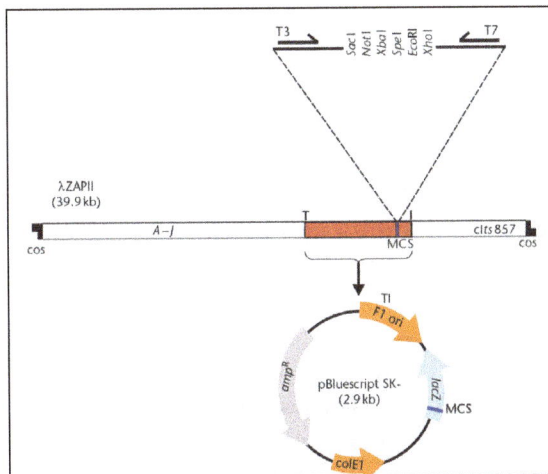

Diagrammatic illustration of the bacteriophage lambda cloning vector λ lZAPII (Stratagene). This 39.9 kb linear DNA molecule can accept inserts up to 10 kb. It has a multicloning site (MCS) (expanded above) flanked by T7 and T3 polymerase promoters to which T7 and T3 primers bind. The MCS is within the lacZ gene allowing blue/white selection of inserts. The region of the vector containing the insert can be excised to the plasmid pBluescript (SK-) as shown, by infection of cells harboring λZAPII with a helper phage. λZAPII also contains terminal cos sites, lambda genes A–J and a temperature-sensitive mutation within the cI repressor that represses lysis.

P1 is another naturally occurring E. coli phage. It has a linear 110 kb genome that upon infection is circularized through the action of the P1 encoded Cre recombinase which catalyzes recombination at loxP recognition sites. Vectors based upon P1 that can maintain inserts of 70–100 kb have been developed. Advantages of P1 vectors include:

- Stability of inserts, because the vector is single copy.

- DNA is present as a closed circle within the cell and so can be isolated using a standard alkaline lysis miniprep.

- Constructs are introduced into the cell by the highly efficient process of packaging and infection.

Bacterial Artificial Chromosomes

BACs are vectors based on the E. coli F factor, which is normally present at one to two copies per cell. They can accept inserts of up to 300 kb and can be directly introduced into deoR strains of E. coli by electroporation, thus avoiding the use of packaging extracts. Their stability and the ease by which clone DNA can be miniprepped for analysis, or end sequencing, make BACs and P1 artificial chromosomes (PACs) (see below) the current vectors of choice for most wholegenome libraries. For a detailed experimental description of BAC library construction. A diagram of a typical BAC vector, pBACe3.6

P1 Artificial Chromosomes

PACs are a hybrid of P1 bacteriophage and BAC vectors. They are low copy number, have good recombinant selection and can be introduced into the cell by electroporation.

Yeast Artificial Chromosomes

Yeast artificial chromosome (YAC) vectors are shuttle vectors containing yeast telomeres, centromere, an origin of replication and selectable markers allowing replication in yeast and E. coli. They can accommodate inserts of up to 2 megabases (Mb). The resulting constructs are transformed into yeast by spheroplast absorption and therein behave like an extra chromosome. Saccharomyces cerevisiae is quite tolerant of extreme base composition and repeats, and so YACs are often useful for cloning sequences that are unclonable in E. coli.

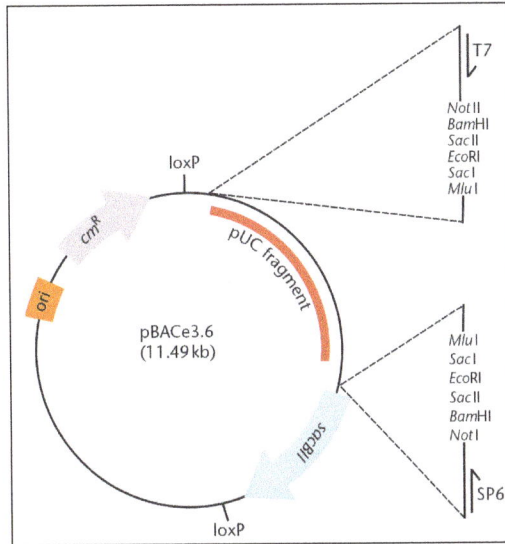

Diagrammatic illustration of the bacterial artificial chromosome (BAC) cloning vector pBACe3.6. This 11.49 kb doublestranded, closed-circular DNA molecule has an origin of replication (ori) allowing single-copy propagation in E. coli, a gene encoding resistance to the antibiotic chloramphenicol (cm^R) and loxP recombination sites that flank the region into which DNA is inserted. Dual multicloning sites (MCS) flank a pUC stuffer fragment. The parent vector possesses this region to allow multicopy growth in E. coli allowing easy preparation of the vector. Restriction of the vector with any of the MCS restriction enzymes excises this fragment. The MCS is flanked by T7 and T3 polymerase promoters, to which T7 and T3 primers bind, and is located between the sacBII gene and its promoter so that inserted DNA disrupts expression of its cytotoxic product, ensuring that only recombinants grow on media containing sucrose.

Human Artificial Chromosomes

With the goal of creating stable transgenic mammalian cell lines, researchers have been developing mammalian artificial chromosomes (MACs) and human artificial chromosomes (HACs). This was achieved in 1997 when Harrington et al. reported the construction and maintenance, in a human cell line, of human artificial chromosomes containing 0.3–2 Mb of inserted a-satellite DNA.

Introduction of Deoxyribonucleic Acid into the Cell

Although some bacteria can spontaneously take up DNA, this is not a widespread occurrence and so a number of methods have been developed for the directed transfer of cloned DNA into the cell.

Transfer into Bacteria

There are four common methods for DNA transfer into bacteria.

Transformation

Transformation is the process of plasmid uptake by bacteria. This process is typically chemically induced with the most common method involving treatment with calcium chloride. Here log phase cells are washed and resuspended in 0.1 M calcium chloride. These cells are incubated on ice with the DNA whereupon a calcium ion–DNA complex is formed on the outside of the cell. If this mixture is subsequently heat-shocked by incubating at 42C for 60–90 s, the cell then takes up this complex. Typically 0.01–0.1% of the DNA is introduced generating up to 108 transformants per microgram of pUC18 DNA.

Electroporation

Application of a brief high-voltage electric shock to a suspension of cells transiently induces the presence of pores in the cell membrane through which DNA can enter the cell. Electroporation can be highly efficient giving more than 10^{10} transformants per microgram of pUC18 DNA.

Transfection

Transfection is defined as the transformation by viral or phage DNA, and can be achieved by either of the previous methods.

Infection

Infection of bacteria can reach efficiencies close to 100%. Typically cells are first grown under conditions that favor induction of the phage receptor; for example E. coli to be infected with phage lambda are cultured in media containing magnesium ions and maltose. Prepared cells can simply be incubated with phage particles, typically at 37 °C for 20–30 min, during which time the phage are absorbed.

Transfer into Eukaryotic Cells

The introduction of DNA into eukaryotic cells is technically more demanding and often less efficient than transfer into bacteria. Nevertheless, a number of methods have been developed involving calcium phosphate-mediated transfection, DEAE dextran transfection, electroporation, microinjection, infection (viral vectors), electroporesis, protoplast/spheroplast fusion and lipofection. The method used is dependent on the type of genetic material being handled, the reason that DNA is being introduced and more critically the type of cell being transformed.

Polymerase Chain Reaction (PCR)

Polymerase Chain Reaction (PCR) is a technique used to make numerous copies of a specific segment of DNA quickly and accurately. The polymerase chain reaction enables investigators to obtain the large quantities of DNA that are required for various experiments and procedures in molecular biology, forensic analysis, evolutionary biology, and medical diagnostics.

The PCR technique is based on the natural processes a cell uses to replicate a new DNA strand. Only a few biological ingredients are needed for PCR. The integral component is the template DNA—i.e., the DNA that contains the region to be copied, such as a gene. As little as one DNA molecule can serve as a template. The only information needed for this fragment to be replicated is the sequence of two short regions of nucleotides (the subunits of DNA) at either end of the region of interest. These two short template sequences must be known so that two primers—short stretches of nucleotides that correspond to the template sequences—can be synthesized. The primers bind, or anneal, to the template at their complementary sites and serve as the starting point for copying. DNA synthesis at one primer is directed toward the other, resulting in replication of the desired intervening sequence. Also needed are free nucleotides used to build the new DNA strands and a DNA polymerase, an enzyme that does the building by sequentially adding on free nucleotides according to the instructions of the template.

Working Principle of Polymerase Chain Reaction

PCR is largely based on a series of 20-40 repeated thermal cycles of heating and cooling to facilitate DNA replication by enzymatic reaction. The amount of target sequence doubles with each thermal cycle which leads to an exponential amplification represented by 2(no. of cycles). Therefore, a PCR thermal cycling of 40 repeats will generate 1,099,511,627,776 copies of DNA from a single copy of the template. The function and purpose of each step in a PCR Reaction.

Initialization

The reaction is heated to 94–96 °C (or 98 °C if extremely thermostable DNA polymerases are used), for 2-10 minutes. This step activates the DNA polymerase and reagents in the reaction as well as denatures other contaminants (if any) in the mixture. In application such as colony screening, small amount of cells can be directly used as template. This initialization step will lyse the cells to release the DNA and denature other cellular proteins including DNases. If a chemical or antibody Hot-Start DNA polymerase is used for PCR, this step is mandatory to heat activate the enzyme.

Denaturation

The reaction is heated to 94–98 °C for 20–30 seconds. This is the first step of thermal cycling. Double-stranded DNA is denatured in this step as hydrogen bonds between the strands are broken due to high temperature, a process also known as DNA melting.

Annealing

Primers are oligonucleotides that can bind to specific sequence of the DNA template to guide DNA polymerase replication. To allow primers to anneal to the single-stranded DNA template, the reaction temperature is typically lowered to 50-65 °C for 20-40 seconds. The optimal temperature for annealing depends on the melting temperature (T_m) of the primers: the temperature at which half of the DNA duplex would dissociate to become single stranded. The primer cannot anneal to the template if the annealing temperature is set too high, and non-specific priming, which leads to non-specific amplification, may happen if the annealing temperature is too low. Thus, the ideal annealing temperature is usually about 3-5 °C below the T_m of the primers, high enough for specific primer annealing, but low enough to allow for efficient priming. The primer concentration in the reaction mixture is usually much higher than that of the DNA template so that primer-template hybridization is greatly favored over re-annealing of the template strands. As soon as the primer anneals to the template, DNA polymerase can start incorporating dNTPs onto the template.

Extension/Elongation

In this step, DNA polymerase binds to the primer/template complex and starts incorporating dNTPs in a 5' to 3' direction on the synthesizing strand. This yields a newly synthesized DNA strand complementary to the template strand. The optimum temperature of extension/elongation varies for different DNA polymerases. The common Taq DNA polymerase (heat resistant) works ideally at 72-78 °C. The duration of extension depends both on the length of the original template, a.k.a. amplicon and the speed of the DNA polymerase. Typical DNA polymerases polymerizes at a speed of 1-1.5 kb/min under its optimal temperature.

Final Elongation

The reaction mixture is kept at 72-78 °C (optimal working temperature for most polymerases) for 5-15 minutes in the last cycle of elongation. This step ensures that any remaining single-stranded DNA is fully extended after the last PCR cycle.

Final Hold

The temperature is lowered to 4-15 °C for an indefinite time for short-term storage of the reaction mixture.

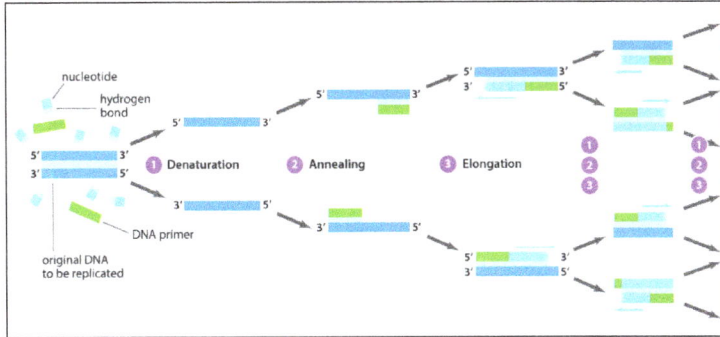

1) Denaturation: the reaction is heated to 94–98 °C for 20–30 seconds to break the hydrogen bonds between the strands. 2) Annealing: The reaction temperature is lowered to 50-65 °C for 20-40 seconds to allow primers to anneal to the template strands. 3) Elongation: the temperature is increased (optimal temperature dependenton DNA Polymerase used e.g. 72-78 °C for Taq Polymerase) to allow for the addition of dNTPs. The amount of target sequence doubles with each thermal cycle which leads to an exponential amplification represented by 2.

To check whether the correct target DNA fragment has been amplified, gel electrophoresis is a quick method to examine the molecular sizes (in bp) of the amplified products. The size(s) of the amplified PCR product is estimated by comparing it to a DNA ladder, a molecular weight marker which contains DNA fragments of known sizes.

Many factors can interfere with a PCR reaction. Some are easy to optimize (e.g. thermal cycling conditions and Mg^{2+} concentration) while others are trickier to manipulate and would require accumulative experience to tackle (e.g. primer design and PCR buffer components/additives). In general, factors that are important in PCR include DNA templates, DNA polymerase, primer design, buffer components, additives and inhibitors and thermal cycling conditions.

PCR - Types of DNA Templates and Strategies

The type of DNA template used for the amplification of a target region can affect the PCR efficiency and accuracy. Below three types of common PCR templates are introduced along with strategies in improving the PCR results.

GC Rich Template

GC content refers to the percentage of guanine and cytosine bases of a DNA region. GC rich template (GC content > 60%) has increased hydrogen bond strength as three hydrogen bonds form between the GC base pair comparing to two between the AT base pair. High GC content leads to increased likelihood of secondary structure generation, which are structures formed when complementary regions of a single stranded DNA molecule loop back and form hydrogen bonds. Both increased hydrogen bond strength and secondary structures hinder template denaturation and primer annealing in PCR. Secondary structures may also lead to pre-mature termination as a result of polymerase arrest. Annealing of primers at alternative binding sites on the template, a.k.a competitive binding is often strong in GC-rich templates which cause incorrect amplicons to be created. Promoter regions of most housekeeping genes, tissue-specific genes and tumor suppressor genes are often rich in GC content, posing difficulties for amplification of these templates.

Strategies to improve the amplification of high GC content templates include optimization of thermal cycling conditions and the use of additives. High annealing temperature can help prevent non-specific binding of primers and amplification. The duration of the annealing time must be optimized to be long enough to allow for the primer to anneal properly to the template, but short enough to prevent competitive binding. Additives such as DMSO and Betaine can act as secondary structure destabilizers, thus lowering the melting temperature and the denaturation duration and promoting specific primer annealing.

AT Rich Templates

AT rich templates often require a lower annealing temperature as the hydrogen bonds between the base pairs are not as strong as GC rich templates. On the other hand, non-specific primer annealing can occur under low annealing temperature.

The use of additives such as TMAC can increase the Tm, thus increasing the primer specificity. The concentration of TMAC added must be optimized as enzyme amplification can be inhibited at high concentrations of this additive. A lower extension temperature such as 65-68 °C for AT rich templates also helps with amplification, as DNA melting at the typical extension temperature of 72 °C prevents successful extension of extremely AT rich regions.

Long Templates

Amplifying long templates in PCR while maintaining high yield, specificity and fidelity is often difficult. Longer templates usually have higher likelihood of DNA template breaks or degradation caused by sample preparation or depurination. Depurination refers to a chemical reaction during which the β-N-glycosidic bond in the purine nucleosides (e.g. A and G) is cleaved to release a nucleic base due to hydrolysis. As Taq DNA polymerase will not extend through apurinic positions during replication, depurination is an important limiting factor of long templates amplification. DNA polymerases sometimes mis-incorporate nucleotides onto the 3' end during replication, which may result in pre-matured termination and accumulation of truncated products. Thus, the amplification efficiency decreases as the amplicon size increases due to the accumulation of truncated products. Several strategies can be employed to avoid this:

- Using higher pH: Lower pH promotes depurination of templates. Increasing pH of a PCR reaction thus protects template from depurination damage, thus decreases the number of truncated products and increases the yield.

- Adding glycerol or DMSO: The processing efficiency of the DNA polymerase becomes important when working with longer templates. DMSO and glycerol are dissociating additives that destabilize DNA and subsequently alters the melting characteristics of the double stranded DNA. In this way, the denaturation and annealing temperatures required are lowered, allowing for faster and more efficient denaturation and annealing steps.

- Decreasing denaturing time: When denaturation duration is long, primer annealing is interfered and depurination of templates occurs. Template quality is crucial for ensuring amplicon quality and this can be enhanced by decreasing the denaturing time.

- Increasing extension time: In long template PCR, the extensions time can be increased to allow enough time for DNA polymerase to add nucldotides. This decreases the number of truncated products synthesized thus increasing the yield of longer PCR products as a result.

- Using a "proofreading" DNA polymerase: A secondary thermo-stable DNA polymerase that possesses a 3'-to 5'-exonuclease, or "proofreading," activity can remove misincorporated nucleotides, which greatly improves the ability of the predominant DNA polymerase to synthesis longer products with higher yield

PCR - Primer Design

Primer design is essential for successful PCR. Below are important considerations for a good primer design which ensures specific annealing and facilitates efficient amplicon extension.

- The optimal primer length is generally between 18-30 nucleotides. Longer primers promote specific annealing but also have higher melting temperature, while shorter primers bind more easily to the template with lower specificity.

- The optimal melting temperature of primers is between 50 °C – 65 °C. The forward and reverse primer pairs should have T_m within 5 °C of each other, so both primers bind simultaneously and efficiently.

- The annealing temperature (T_a) should be set to be no more than 5 °C lower than the T_m, high enough to promote specific binding and low enough to allow efficient annealing.

- Design primers with GC content between 40%-60% to promote specific annealing. The GC content of the target sequence greatly affects the melting temperature, as higher GC content sequence has a higher melting temperature.

- Ending the primer 3' terminal with Gs or Cs can promote specific target binding. However, there should be no more than 3 Gs or Cs at the last five bases.

- Target sequences containing secondary structures may result in poor primer annealing and low yield, and should be avoided if possible.

- Avoid having repeats of 4 or more nucleotide, or dinucleotide repeats (e.g. ATATATAT).

- Avoid having complementary sequences within the primer or between the forward and reverse primers which usually lead to self-dimers or primer-dimers. As the primer concentration is much larger than template concentration in a reaction, the formation of self-dimers and primer-dimers will seriously compete with template-primer annealing and affects product yield.

PCR - Buffer Components

The most common components of PCR buffers are Tris-HCl, Magnesium ions, KCl or other salt solutions, and dNTPs. Co-solvents or additives may also be included to enhance the amplification. Buffer components greatly affect PCR result in means of pH, cofactors, salt concentration and ionic strength. Thermostable DNA polymerases require a relative stable pH to perform with optimized activity and fidelity. Tris-base buffers are generally use due to their stabilization of pH over a broad range of temperatures and changes in reaction mix composition. Magnesium is a co-factor of thermostable DNA polymerases and its presence in the reaction is essential for PCR success. For example, the mostly commonly used Taq polymerase activity drops dramatically with insufficient amount of free Mg^{2+}, while the enzyme fidelity and amplification specificity is reduced by excess amount of the ion. The performance of Pfu DNA polymerase, on the other hand, is less dependent on magnesium concentration. Thus, Magnesium-free reaction buffer is often supplied with many DNA polymerases with a separate aliquot of Mg ion solution for researches to optimize each reaction using varying Mg^{2+} concentration. Salt can neutralize the negative charges on the phosphate backbone of DNA, thus reducing the repulsion between two DNA strands. Some salt in PCR buffer can stabilize primer-template binding during annealing to help polymerases start extension. It is found that some level of KCl in PCR buffer can greatly increase DNA polymerase activity and the yield of short length products, while longer templates are more efficiently amplified at lower concentration of K^+. The use of KCl and $(NH_4)_2SO_4$ in combination have shown to destabilize hydrogen bonds between mismatched bases, thus promoting specificity of primer annealing. dNTP (deoxynucleoside triphosphate) are building blocks for the DNA polymerases to incorporate onto the synthesis strand. A balanced mix of dATP, dGTP, dCTP and dTTP is necessary for best incorporation. However, an unbalanced dNTP concentration can sometimes be used to promote DNA polymerases misincorporation in certain applications such as random mutagenesis studies.

PCR buffer usually contain various additives to enhance amplification.

PCR - Additives and Inhibitors

To increase the yield, specificity and consistency of PCR reactions, PCR enhancing agents or additives can be added. Common additives include DMSO, Betaine, Formamide, TMAC, BSA, non-ionic detergents, glycerol, and etc. They are especially important in enhancing PCR amplification of long or GC rich templates. The use of additives needs to be optimized based on experiment conditions, purposes and often requires repeated trials and analysis.

Table: Examples of PCR additives and their mechanisms of action.

Additives	Mechanism of Action
Dimethylsulfoxide (DMSO)	• Destablize secondary structures, • Promote dissociation of DNA strands, • Prevent primer/target mis-priming.
Betaine	• Destablize secondary structures, • Promote dissociation of DNA strands, • Protect DNA polymearase from denaturation, • Remove inhibitory products accumulated during amplification.
Glycerol	• Enhancing hydrophobic interactions between protein domains, • Promote dissociation of DNA strands.
Formamide	• Destabilize mis-matched bonds and base pairing interactions
Bovine Serum Albumin (BSA)	• Reducing the inhibitory effects of inhibitors present in samples (e.g. humic acids in blood sample), • Stabilize the enzymatic activity of DNA polymerases, • Enhance the effects of organic solvents.
Dithiothreitol (DTT)	• Maintain active conformational structures of DNA polymerases.
Tetramethyl Ammonium Chloride (TMAC)	• Reduce potential DNA mismatch and improve the stringency of hybridization reactions.
Gelatin	• Stabilize DNA polymerases.
Non-Ionic Detergents (Tween-20, Triton X-100)	• Overcome inhibitory effects of strong ionic detergents, • Stablize DNA polymerases, • Dissociate secondary structures.
Ammonium Ions	• Render reaction more tolerant of non-optimal conditions.
Tetramethylene Sulphoxide	• Promote dissociation of DNA strands.

PCR is also prone to inhibiting substances present in samples. Some residual inhibitors are present in specific types of DNA samples and this will require sample-specific nucleic acid isolation protocols prior to PCR. PCR inhibitors can also be introduced during sample preparation. Inhibitors may degrade or modify the DNA template, disturb annealing of the primers to the DNA template due to competitive binding of the inhibitor to the template, and degrade, inhibit or alter DNA polymerase activity. Refer to table for examples of PCR inhibitors and their mechanism of action.

Table: Examples of PCR inhibitors and their mechanisms of action.

Inhibitor	Mechanism of Action
Polysaccharides	• Co-percipitation with nucleic acid. • Reduction in the ability to resuspend precipitated RNA.
Bacterial Cells	• Degradation/sequestration of nucleic acids.
Melanin	• Cross-linking with nucleic acids. • Change of chemical properties of nucleic acids.
Collagen	• Binding to nucleic acid and inhibition of enzymes.
Humic Acid	• Binding/adsorption to nucleic acid and enzymes.
Haematin	• Incomplete melting of DNA and inhibition of enzymes.
Metal Ions	• Reduction in specificity of primers.
Detergents	• Degradation of polymerases.
Calcium	• Inhibition of DNA polymearse or reverse transcriptase activity. • Competition with co-factors of the polymerase.
Tannic Acid	• Chelation of metal ions including Mg^{2+}. • Binding DNA polymerase.
EDTA	• Chelation of metal ions including Mg^{2+}.
Antiviral	• Competition with nucleotides. • inhibition of DNA elongation.
Exogenic DNA	• Competition with template.

PCR - Thermal Cycling Conditions

The lengths and temperatures for the thermal cycle in PCR can be modified to promote amplification under different experiment conditions. In some cases, the denaturation cycle can be shortened or a lower temperature used to reduce the chances of nucleotides depurination, which can lead to truncated products or mutations in the PCR products.

Annealing temperature and extension time are the most commonly altered parameters to reduce nonspecific amplification and primer-dimer formation. The optimal annealing temperature (Ta) is often within 5 °C of the Tm of the primers. Higher Ta leads to increased annealing stringency and specificity, while lower Ta results in more efficient annealing and higher yields. The extension time should be optimized based on the size of the PCR product and the DNA polymerase being used. In general, it takes minimally

1 minute per kb of amplicon for non-proofreading DNA polymerases and 2 minutes per kb of amplicon for proofreading DNA polymerases. Excessively long extension times allow the intrinsic $5' \to 3'$ exonuclease activity of Taq DNA polymerase to generate undesired products, so should be avoided.

Typically 25–35 thermal cycles in PCR can synthesize enough products for downstream application. As the number of cycles increases, the risk of undesirable PCR products appearing in the reaction increases, so it is important to set an appropriate number of thermal cycles in PCR.

Touchdown PCR is one method of PCR with specially modified thermal cycling conditions to enhance PCR results. In the initial cycles of touchdown PCR, the Ta (annealing temperature) is set to be high, usually several degrees above the estimated Tm of primers, to only tolerate specific primer binding to the template, thus usually amplifying the correct target sequence. Ta is then gradually decreased for every subsequent cycles determined by the scientists (e.g. 1-2 °C/every second cycle), to allow more efficient primer-binding and amplification. Since the desired products were already amplified during the earlier cycles, they will out compete the nonspecific sequence the primer might bind to at lower Ta. Touchdown PCR therefore offers a simple and rapid method to optimize PCR with enhanced specificity, efficiency and yield, and is particularly useful to amplify difficult templates.

Nucleic Acid Hybridization

The DNA double helix consists of two strands of nucleotides twisted around each other and held together by hydrogen bonding between the bases. If a solution of DNA is heated, the input of energy makes the molecules vibrate and the hydrogen bonds start coming apart.

If the temperature is high enough, the DNA comes completely apart into two separate strands. This is known as denaturation or "melting". Since the GC base pair has three hydrogen bonds compared to two holding AT together, GC base pairs are stronger than AT base pairs. Therefore, as the temperature rises, AT pairs come apart first and regions of DNA with lots of GC base pairs melt at higher temperatures.

The melting temperature of a DNA molecule is defined as the temperature at halfway point on melting curve. The halfway point is used because it is more accurate than trying to guess precisely where melting is complete.

Melting is followed by measuring the UV absorption, since disordered DNA absorbs more UV light. Overall, the higher the proportion of GC base pairs, the higher the melting temperature of a DNA molecule.

If denatured DNA is cooled again, the single DNA strand will recognize partners by base pairing and double stranded DNA will re-form. This is referred to as annealing. For proper annealing, DNA must be cooled slowly to allow time for single strand to find the correct partners.

Consider two completely different. DNA molecules. If they are mixed, melted and then cooled to re-anneal the single strand, each single strand will recognize and pair with its original complementary strand.

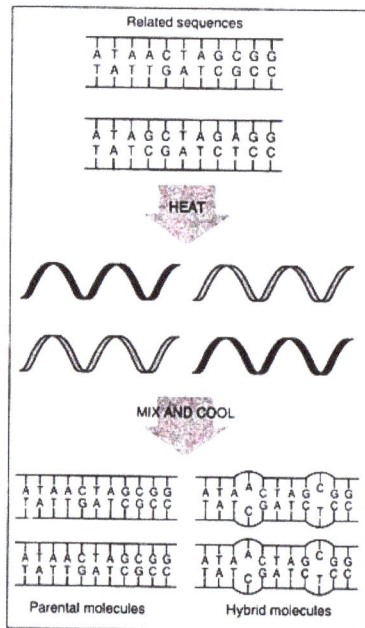

Figure: Hybrid DNA formed by Melting and Re-Annealing.

Suppose on the other hand, two closely related DNA molecules are used. Although the sequences may not match perfectly, nonetheless, if they are similar enough, some base pairing will occur. The result will be formation of hybrid DNA molecules.

The followings are uses of Nucleic acid hybridization:

(a) To Test the Frequency of Relatedness between Two DNA Molecules:

To do this, a sample of first DNA molecule is heated to melt it into single-stranded DNA. The single strands are then attached to suitable filter. Next, the filter is treated chemically to block any remaining sites that would bind DNA. Then, after melting, a solution of second DNA molecule is poured through the filter.

Some of the single strands of DNA molecule No. 2 will base pair with the single strands of DNA molecule No. 1 and will stick to the filter. The more closely related the two molecules are, the more hybrid molecules will be formed and the higher the proportion of molecule No. 2 will bound by the filter.

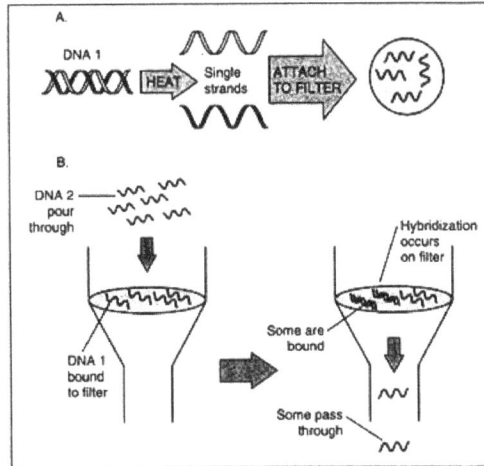

Figure: Relatedness of DNA by Filter Hybridization.

For example, if the DNA for a human gene, such as haemoglobin, is fully melted and bound to a filter, then DNA for the same gene but from different animals could be tested. We might expect Gorilla DNA to bind strongly, frog DNA to bind weakly and mouse DNA to be intermediate.

(b) To Isolate Genes for the Process of Cloning:

Suppose we already have the human haemoglobin gene and want to isolate the corresponding Gorilla gene. First, the human DNA is bound to the filter as before. Then gorilla DNA is cut into short segments with a suitable restriction enzyme.

The gorilla DNA is heated to be melted into single strands and poured over the filter. The DNA fragment that carries the gorilla gene for haemoglobin will bind to human haemoglobin gene and remain stuck to the filter. Other unrelated genes will not hybridize. This approach allows isolation of new genes provided a related gene is available for hybridization.

(c) To Search for Identical or Similar Sequences in Experimental Sample of Target Molecules:

A wide range of method based on hybridization is used for analysis in molecular biology. The basic idea in each case is that a known DNA sequence acts as a "probe". Generally, the probe molecule is labelled by radioactivity or fluorescence for ease of detection.

The probe is used to search for identical or similar sequences in experimental sample of target molecules. Both the probe and target DNA must be treated to give single-stranded DNA molecules that can hybridize to each other by base pairing. In previous example, the probe DNA would be the human haemoglobin DNA since the sequence is already known. The gorilla DNA would be sample of target molecules.

Methods of Nucleic Acid Hybridization

Southern Hybridization

The basic principle behind the southern hybridization is the nucleic acid hybridization. Southern hybridization commonly known as southern blot is a technique employed for detection of a specific DNA sequence in DNA samples that are complementary to a given RNA or DNA sequence. It was the first blotting technique to be devised, named after its pioneer E.M Southern, a British biologist. Southern blotting involves separation of restricted DNA fragments by electrophoresis and then transferred to a nitrocellulose or a nylon membrane, followed by detection of the fragment using probe hybridization.

Separated by electrophoresis is transferred from gel to a membrane which in turn is used as a substrate for hybridization analysis employing labeled DNA or RNA probes specific to target fragments in the blotted DNA. Southern hybridization helps to detect specific fragment against a background of many other restriction fragments. Southern blotting is a technique which is used to confirm the identity of a cloned fragment or for recognition of a sub-fragment of interest from within the cloned DNA, or a genomic DNA. Southern blotting is a prerequisite to techniques such as restriction fragment length polymorphism (RFLP) analysis.

Procedure

- The high-molecular-weight DNA strands are fractioned using restriction enzymes.

- The DNA fragments are separated based on size by agarose gel electrophoresis.

- The gel with the restricted fragments is then laid on a filter paper wick which serves as a connection between the membrane and the high salt buffer.

- The nitrocellulose membrane is placed on top of the gel and a tower of filter papers is used to cover it and these are kept in place with a weight. The capillary action drives the buffer soaking through the filter paper wick, through the gel and the membrane and into the paper towels. Along with the buffer passing through the gel the DNA fragments are also carried with it into the membrane and they bind to the membrane. This causes an effective transfer of fragments (up to 15 kb in length taking around 18hours or overnight.

For DNA fragments larger than 15 kb, before blotting an acid such as diluted HCl is used to treat the gel that depurinates the DNA fragments causing breakage of DNA into smaller pieces, resulting in more efficient transfer from the gel to membrane.

Now a days blotting is also done by applying electric field. This electro blotting technique depends upon current and transfer buffer solution to nucleic acids onto a membrane. Following electrophoresis, a standard tank or semi-dry blotting transfer system is set up. A stack is put together in the following order from cathode to anode: Sponge, three sheets of filter paper soaked in transfer buffer gel, PVDF or nitrocellulose membrane, three sheets of filter paper soaked in transfer buffer and then again sponge. Importantly the membrane should be located between the gel and the positively-charged anode, as the current and sample will be moving in that direction. Once the stack is prepared, it is placed in the transfer system, and suitable current is applied for a specific period of time according to the materials being used.

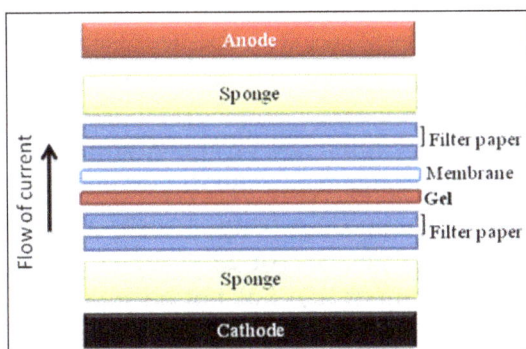

Figure: Set up for Electro-blotting system.

- For using alkaline transfer methods, the DNA gel is placed into an alkaline solution (like that of sodium hydroxide) causing denaturation of the double-stranded DNA. Denaturation in an alkaline environment enhances the binding between the negatively charged DNA and the positively charged membrane, causing separation to single DNA strands for further hybridization to the probe, alongside destroying any residual RNA that may persist in DNA. The membrane is washed with buffer to remove unbound DNA fragments.

- The membrane which contains the transferred fragments is heated in presence or absence of vacuum at 80°C for 2 hours or exposed to ultraviolet radiation (nylon membrane) for permanent attachment of the transferred DNA to the membrane.

- The obtained membrane is then hybridized with a probe (a DNA fragment with a specific sequence whose presence in the target DNA is to be determined).

- Labeling of the probe DNA is done for easy detection, usually radioactivity is incorporated or the molecule is tagged with a fluorescent or chromogenic dye. The hybridization probe may be made of RNA, instead of DNA in some cases where the target is RNA specific.

- Washing of the excess probe from the membrane is done by using saline sodium citrate(SSC buffer) after the hybridization step and the hybridization pattern is studied on an X-ray film by autoradiography (for a radioactive or fluorescent probe), or via color development on membrane if a chromogenic detection method is employed.

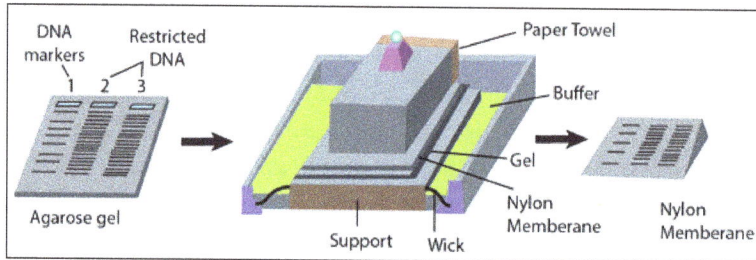

Figure: Steps of Southern Hybridization.

Analysis of Southern Blot

Hybridization of the probe to a specific DNA fragment on the membrane indicates the presence of a complementary fragment in the DNA sequence. Southern hybridization performed by digestion of genomic DNA using a restriction enzyme digestion, helps in determining the number of sequences (or gene copies) in the genome. For a probe hybridizing to a single DNA segment that has not been cut by the restriction enzyme, a single band is observed and on the other hand multiple bands will likely be observed when the hybridization occurs between the probe and several highly similar target sequence (Due to sequence duplication). Alterations in the hybridization conditions like enhancing the hybridization temperature or decreasing salt concentration, helps in altering specificity and hybridization of the probe to sequences that are less than 100% similar.

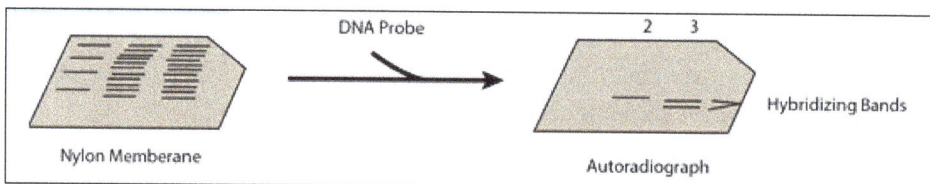

Figure: Southern hybridization analysis.

Applications

Southern blotting has been exploited for various applications which include:

- Clone identification: One of the most common applications of Southern blotting is identification and cloning of a specific gene of interest. Southern blotting is carried out for identification of one or more restriction fragments that contain the gene of interest in genomic DNA.. After cloning and tentative identification of the desired recombinant by employing colony or plaque hybridization, southern blotting is further is used to confirm the clone identification and possibly to locate a shorter restriction fragment, containing the sequence of interest.

- Restriction fragment length polymorphism Analysis: Another major application of Southern hybridization is restriction fragment length polymorphism (RFLP) mapping, which is crucial in construction of genome maps.

Northern Hybridization

Northern blotting was developed by James Alwine, George Stark and David Kemp. Northern blotting drives its name because of its similarity to the first blotting technique, which is Southern blotting, named after the biologist Edwin Southern. The major difference is that RNA being analyzed rather than DNA in the northern blot.

Expression of a particular gene can be detected by estimating the corresponding mRNA by Northern blotting. Northern blotting is a technique where RNA fragments are separated by electrophoresis and immobilized on a paper sheet.Identification of a specific RNA is then done by hybridization using a labeled nucleic acid probe. It helps to study gene expression by detection of RNA (or isolated mRNA) in a sample.

In Northern blotting, probes formed of nucleic acids with a sequence which is complementary to the sequence or to a part of the RNA of interest. The probe can be DNA, RNA or chemically synthesized oligonucleotides of minimum 25 complementary bases to the target sequence.

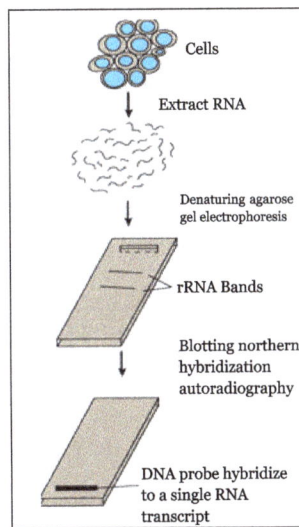

Figure: Steps of Northern Hybridization.

Procedure

The northern blotting involves the following steps:

- Total RNA is extracted from a homogenized tissue sample or cells. Further eukaryotic mRNA can then be isolated by using of oligo (dT) cellulose chromatography to isolate only those RNAs by making use of a poly A tail.

- The isolated RNA is then separated by gel electrophoresis.

- The RNA samples separated on the basis of size are transferred to a nylon membrane employing a capillary or vacuum based system for blotting.

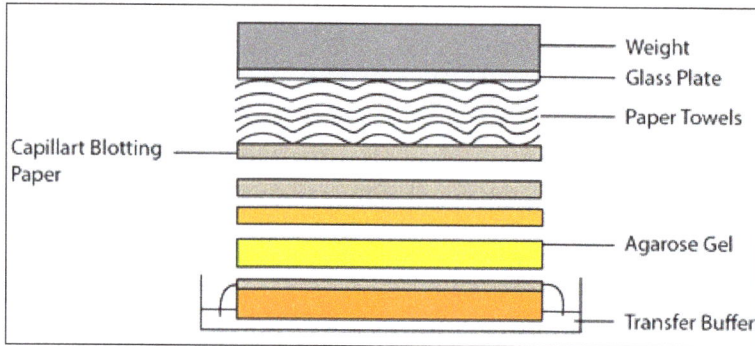

Figure: Setup for Northern blotting.

- Similar to Southern blotting, the membrane filter is revealed to a labeled DNA probe that is complementary to the gene of interest and binds.

- The labeled filter is then subjected to autoradiography for detection.

- The net amount of a specific RNA in a sample can be estimated by using Northern blot. This technique is widely used for comparing the amounts of a particular mRNA in cells under different conditions. The separation of RNA samples is often done on agarose gels containing formaldehyde as a denaturing agent as it limits the RNA to form secondary structure.

Analysis of Northern Blot

RNA extract is electrophoresed in an agarose gel, using a denaturing electrophoresis buffer (containing formaldehyde) to ensure that the RNAs do not form inter- or intra-molecular base pairs, as base pairing would affect the rate at which the molecules migrate through the gel. After electrophoresis, the gel is blotted onto a nylon or nitrocellulose membrane, and hybridized with a labeled probe. If the probe is a cloned gene, the band that appears in the autoradiograph is the transcript of that gene. The size of the transcript can be determined from its position within the gel, and if RNA from different tissues is run in different lanes of the gel, then the possibility of differentially expressed gene can be examined.

Applications

Northern blotting helps in studying gene expression pattern of various tissues, organs, developmental stages, pathogen infection, and also over the course of treatment. It has been employed to study overexpression of oncogenes and down-regulation of tumor-suppressor genes in cancerous cells on comparison with healthy tissue, and also for gene expression of immune-rejection of transplanted organ.

The examination of the patterns of gene expressions obtained under given conditions can help determine the function of that gene.

Northern blotting is also used for the analysis of alternate spliced products of same gene or repetitive sequence motif by investigating the various sized RNA of the gene. This is done when only probe type with variation in one location is used to bind to the target RNA molecule.

Variations in size of a gene product may also help to identify deletions or errors in transcript processing, by altering the probe target that can be used along the known sequence and make it possible to determine the missing region of the RNA.

Colony Hybridization

It is a rapid method of isolating a colony containing a plasmid harboring a particular sequence or a gene from a mixed population. The colonies to be screened are first replica-plated on to a nitrocellulose filter disc that has been placed on the surface of an agar plate prior to inoculation. Master plate is retained for reference set of colonies. The filter bearing the colonies is removed and treated with alkali so that the bacterial colonies are lysed and the DNA they contain is denatured. The filter is then treated with proteinase K to remove protein and leave denatured DNA bound to the nitrocellulose. The DNA is fixed firmly by baking the filter at 80 °C. A labeled probe is hybridized to this DNA which is monitored by autoradiography. A colony whose DNA print gives a positive auto radiographic result on X-ray film can then be picked from the reference plate.

Colony hybridization can be used to screen plasmid or cosmid based libraries.

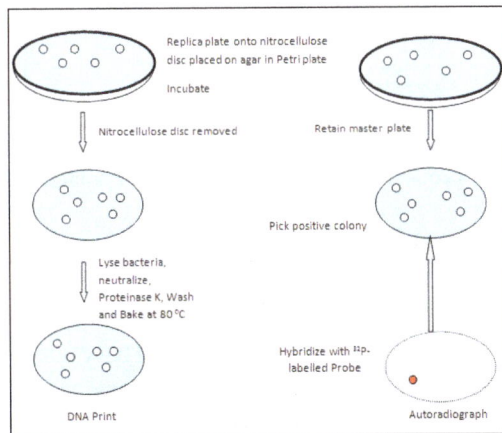

Figure: Colony hybridization.

In Situ Hybridization (ISH)

It is a technique that employs a labeled complementary nucleotide strand (i.e. probe) for localizing specific DNA or RNA sequence targets within fixed tissues and cells (i.e in situ). Probes used for hybridization can be double-stranded DNA probes, single-stranded DNA probes, RNA probes, synthetic oligonucleotides. There are two ways available to detect DNA or RNA targets.

Chromogenic (CISH) in situ hybridization and Fluorescence in situ hybridization (FISH).

Chromogenic In Situ Hybridization (CISH)

It uses the labeling reactions involving alkaline phosphatase or peroxidase reactions to visualize the sample using bright-field microscopy. It is primarily used in molecular pathology diagnostics. CISH can also be employed for samples like fixed cells or tissues, blood or bone marrow smears and metaphase chromosome spreads.

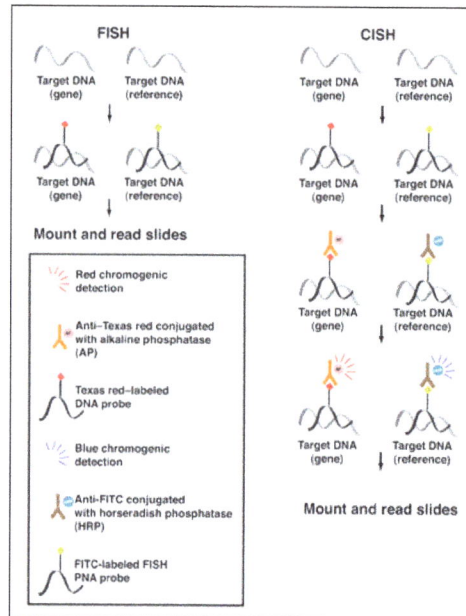

Figure: Use of dual-color Chromogenic in situ hybridization (CISH) in combination with fluorescence in situ hybridization (FISH) probes. FITC, fluorescein isothiocyanate; PNA, peptide nucleic acid.

Fluorescence In Situ Hybridization (FISH)

FISH is a cytogenetic technique that uses fluorescent probes that bind to complementary targets and sample is visualized using epi-fluorescence or confocal microscopy. Using differently labeled probes, we can visualize several targets in a single sample. It is used for spatial-temporal patterns of gene expression and resolving genetic elements in chromosomal preparations.

Cells or tissues to be analyzed are fixed and permeabilized with Proteinase K to allow target accessibility. Probe is constructed and tagged using non-radioactive labels like biotin, digoxigenin or fluorescent dye (FISH). Probe must be large enough to hybridize specifically with its target. Probe is applied to fixed sample and incubated to several hours to allow hybridization. Washing is done to remove non-specific or unbound probe removal. Results are then visualized using either bright-field or confocal microscopy.

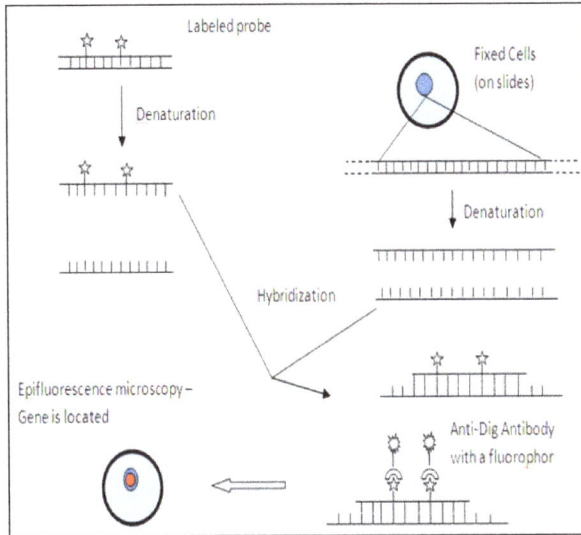

Figure: Fluorescent in-situ hybridization.

Dot Blot and Slot Blot Hybridization

These two techniques represents the simplification of Southern and Western blots saving the time involved in procedures of chromatography, electrophoresis, restriction digestion and blotting of DNA or proteins from the gel to membrane. Here nucleic acid mixture is directly applied (blotted) on to the nylon or nitrocellulose membrane where hybridization between probe and target takes place, denatured to single-stranded form and baked at 80 °C to bind DNA target to membrane. In dot-blot, target is blotted as circular blots whereas in slot-blots, it is in the form of rectangular blots. Due to this, slot-blot offers greater precision in observing different hybridization signals. After blotting, membrane is allowed to dry and non-specific sites are blocked by soaking in blocking buffer containing BSA. It is then followed by hybridization of labeled probe for detection of specific sequences or gene.

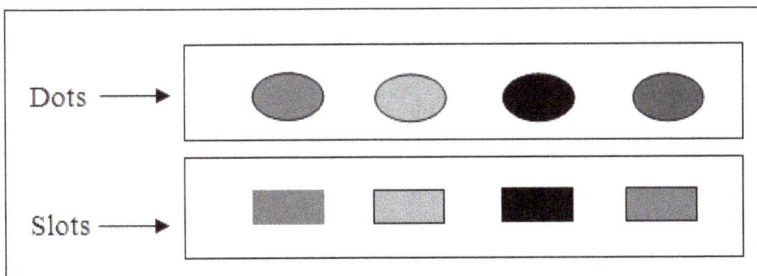

Figure: Dots and slots in dot and slot blot hybridization.

These procedures can only detect presence and absence of particular sequence or gene. It cannot distinguish between two molecules of different sizes as they appear as single dot on membrane. It also has application in detecting alleles that differ in single nucleotide with the help of allele-specific oligonucleotides.

DNA Repair

DNA repair is any of several mechanisms by which a cell maintains the integrity of its genetic code. DNA repair ensures the survival of a species by enabling parental DNA to be inherited as faithfully as possible by offspring. It also preserves the health of an individual. Mutations in the genetic code can lead to cancer and other genetic diseases.

Successful DNA replication requires that the two purine bases, adenine (A) and guanine (G), pair with their pyrimidine counterparts, thymine (T) and cytosine (C). Different types of damage, however, can prevent correct base pairing, among them spontaneous mutations, replication errors, and chemical modification. Spontaneous mutations occur when DNA bases react with their environment, such as when water hydrolyzes a base and changes its structure, causing it to pair with an incorrect base. Replication errors are minimized when the DNA replication machinery "proofreads" its own synthesis, but sometimes mismatched base pairs escape proofreading. Chemical agents modify bases and interfere with DNA replication. Nitrosamines, which are found in products such as beer and pickled foods, can cause DNA alkylation (the addition of an alkyl group). Oxidizing agents and ionizing radiation create free radicals in the cell that oxidize bases, especially guanine. Ultraviolet (UV) rays can result in the production of damaging free radicals and can fuse adjacent pyrimidines, creating pyrimidine dimers that prevent DNA replication. Ionizing radiation and certain drugs, such as the chemotherapeutic agent bleomycin, can also block replication, by creating double-strand breaks in the DNA. (These agents can also create single-strand breaks, though this form of damage often is easier for cells to overcome.) Base analogs and intercalating agents can cause abnormal insertions and deletions in the sequence.

There are three types of repair mechanisms: direct reversal of the damage, excision repair, and postreplication repair. Direct reversal repair is specific to the damage. For example, in a process called photoreactivation, pyrimidine bases fused by UV light are separated by DNA photolyase (a light-driven enzyme). For direct reversal of alkylation events, a DNA methyltransferase or DNA glycosylase detects and removes the alkyl group. Excision repair can be specific or nonspecific. In base excision repair, DNA glycosylases specifically identify and remove the mismatched base. In nucleotide excision repair, the repair machinery recognizes a wide array of distortions in the double helix caused by mismatched bases; in this form of repair, the entire distorted region is excised. Postreplication repair occurs downstream of the lesion, because replication is blocked at the actual site of damage. In order for replication to occur, short segments of DNA called Okazaki fragments are synthesized. The gap left at the damaged site is filled in through recombination repair, which uses the sequence from an undamaged sister chromosome to repair the damaged one, or through error-prone repair, which uses the damaged strand as a sequence template. Error-prone repair tends to be inaccurate and subject to mutation.

Often when DNA is damaged, the cell chooses to replicate over the lesion instead of waiting for repair (translesion synthesis). Although this may lead to mutations, it is preferable to a complete halt in DNA replication, which leads to cell death. On the other hand, the importance of proper DNA repair is highlighted when repair fails. The oxidation of guanine by free radicals leads to G-T transversion, one of the most common mutations in human cancer.

Hereditary nonpolyposis colorectal cancer results from a mutation in the MSH2 and MLH1 proteins, which repair mismatches during replication. Xeroderma pigmentosum (XP) is another condition that results from failed DNA repair. Patients with XP are highly sensitive to light, exhibit premature skin aging, and are prone to malignant skin tumours because the XP proteins, many of which mediate nucleotide excision repair, can no longer function.

DNA Mismatch Repair

DNA mismatch repair (MMR) is a system for recognizing and repairing erroneous insertion, deletion, and mis-incorporation of bases that can arise during DNA replication and recombination, as well as repairing some forms of DNA damage.

Mismatch repair is strand-specific. During DNA synthesis the newly synthesised (daughter) strand will commonly include errors. In order to begin repair, the mismatch repair machinery distinguishes the newly synthesised strand from the template (parental). In gram-negative bacteria, transient hemimethylation distinguishes the strands (the parental is methylated and daughter is not). However, in other prokaryotes and eukaryotes, the exact mechanism is not clear. It is suspected that, in eukaryotes, newly synthesized lagging-strand DNA transiently contains nicks (before being sealed by DNA ligase) and provides a signal that directs mismatch proofreading systems to the appropriate strand. This implies that these nicks must be present in the leading strand, and evidence for this has recently been found. Recent work has shown that nicks are sites for RFC-dependent loading of the replication sliding clamp PCNA, in an orientation-specific manner, such that one face of the donut-shape protein is juxtaposed toward the 3'-OH end at the nick. Loaded PCNA then directs the action of the MutLalpha endonuclease to the daughter strand in the presence of a mismatch and MutSalpha or MutSbeta.

Any mutational event that disrupts the superhelical structure of DNA carries with it the potential to compromise the genetic stability of a cell. The fact that the damage detection and repair systems are as complex as the replication machinery itself highlights the importance evolution has attached to DNA fidelity.

Examples of mismatched bases include a G/T or A/C pairing. Mismatches are commonly due to tautomerization of bases during DNA replication. The damage is repaired by recognition of the deformity caused by the mismatch, determining the template and non-template strand, and excising the wrongly incorporated base and replacing it with the correct nucleotide. The removal process involves more than just the mismatched nucleotide itself. A few or up to thousands of base pairs of the newly synthesized DNA strand can be removed.

Mismatch Repair Proteins

Mismatch repair is a highly conserved process from prokaryotes to eukaryotes. The first evidence for mismatch repair was obtained from *S. pneumoniae* (the hexA and hexB genes). Subsequent work on *E. coli* has identified a number of genes that, when mutationally inactivated, cause hypermutable strains. The gene products are, therefore, called the "Mut" proteins, and are the major active components of the mismatch repair system. Three of these proteins are essential in detecting the mismatch and directing repair machinery to it: MutS, MutH and MutL (MutS is a homologue of HexA and MutL of HexB).

DNA Mismatch Repair Protein, C-2

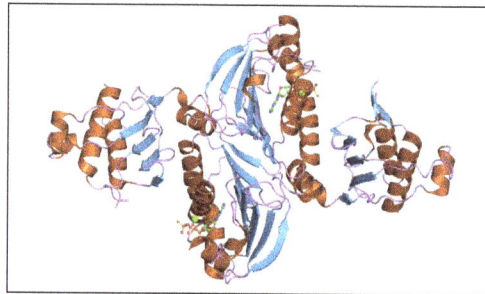

hpms2-atpgs

Identifiers	
Symbol	DNA_mis_repair
Pfam	PF01119
Pfam clan	CL0329
InterPro	IPR013507
PROSITE	PDOC00057
SCOPe	1bkn / SUPFAM
Available protein structures:	
Pfam	structures / ECOD
PDB	RCSB PDB; PDBe; PDBj
PDBsum	structure summary

MutS forms a dimer (MutS$_2$) that recognises the mismatched base on the daughter strand and binds the mutated DNA. MutH binds at hemimethylated sites along the daughter DNA, but its action is latent, being activated only upon contact by a MutL dimer (MutL$_2$), which binds the MutS-DNA complex and acts as a mediator between MutS$_2$ and MutH, activating the latter. The DNA is looped out to search for the nearest d(GATC) methylation site to the mismatch, which could be up to 1 kb away. Upon activation by the MutS-DNA complex, MutH nicks the daughter strand near the hemimethylated site. MutL recruits UvrD helicase (DNA Helicase II) to separate the two strands with a specific 3' to 5' polarity. The entire MutSHL complex then slides along the DNA in the direction of the mismatch, liberating the strand to be excised as it goes. An exonuclease trails the complex and digests the ss-DNA tail. The exonuclease recruited is dependent on which side of the mismatch MutH incises the strand – 5' or 3'. If the nick made by MutH is on the 5' end of the mismatch, either RecJ or ExoVII (both 5' to 3' exonucleases) is used. If, however, the nick is on the 3' end of the mismatch, ExoI (a 3' to 5' enzyme) is used.

The entire process ends past the mismatch site - i.e., both the site itself and its surrounding nucleotides are fully excised. The single-strand gap created by the exonuclease can then be repaired by DNA Polymerase III (assisted by single-strand-binding protein), which uses the other strand as a template, and finally sealed by DNA ligase. DNA methylase then rapidly methylates the daughter strand.

MutS Homologs

When bound, the $MutS_2$ dimer bends the DNA helix and shields approximately 20 base pairs. It has weak ATPase activity, and binding of ATP leads to the formation of tertiary structures on the surface of the molecule. The crystal structure of MutS reveals that it is exceptionally asymmetric, and, while its active conformation is a dimer, only one of the two halves interacts with the mismatch site.

In eukaryotes, MutS homologs form two major heterodimers: Msh2/Msh6 (MutSα) and Msh2/Msh3 (MutSβ). The MutSα pathway is involved primarily in base substitution and small-loop mismatch repair. The MutSβ pathway is also involved in small-loop repair, in addition to large-loop (~10 nucleotide loops) repair. However, MutSβ does not repair base substitutions.

MutL Homologs

MutL also has weak ATPase activity (it uses ATP for purposes of movement). It forms a complex with MutS and MutH, increasing the MutS footprint on the DNA.

However, the processivity (the distance the enzyme can move along the DNA before dissociating) of UvrD is only ~40–50 bp. Because the distance between the nick created by MutH and the mismatch can average ~600 bp, if there is not another UvrD loaded the unwound section is then free to re-anneal to its complementary strand, forcing the process to start over. However, when assisted by MutL, the *rate* of UvrD loading is greatly increased. While the processivity (and ATP utilisation) of the individual UvrD molecules remains the same, the total effect on the DNA is boosted considerably; the DNA has no chance to re-anneal, as each UvrD unwinds 40-50 bp of DNA, dissociates, and then is immediately replaced by another UvrD, repeating the process. This exposes large sections of DNA to exonuclease digestion, allowing for quick excision (and later replacement) of the incorrect DNA.

Eukaryotes have five MutL homologs designated as MLH1, MLH2, MLH3, PMS1, and PMS2. They form heterodimers that mimic MutL in *E. coli*. The human homologs of prokaryotic MutL form three complexes referred to as MutLα, MutLβ, and MutLγ. The MutLα complex is made of MLH1 and PMS2 subunits, the MutLβ heterodimer is made of MLH1 and PMS1, whereas MutLγ is made of MLH1 and MLH3. MutLα acts as an endonuclease that introduces strand breaks in the daughter strand upon activation by mismatch and other required proteins, MutSα and PCNA. These strand interruptions serve as entry points for an exonuclease activity that removes mismatched DNA. Roles played by MutLβ and MutLγ in mismatch repair are less-understood.

MutH: An Endonuclease Present in E. Coli and Salmonella

MutH is a very weak endonuclease that is activated once bound to MutL (which itself is bound to MutS). It nicks unmethylated DNA and the unmethylated strand of hemimethylated DNA but does not nick fully methylated DNA. Experiments have shown that mismatch repair is random if neither strand is methylated. These behaviours led to the proposal that MutH determines which strand contains the mismatch. MutH has no eukaryotic homolog. Its endonuclease function is taken up by MutL homologs, which have some specialized 5'-3' exonuclease activity. The strand bias for removing mismatches from the newly synthesized daughter strand in eukaryotes may be provided by the free 3' ends of Okazaki fragments in the new strand created during replication.

PCNA β-sliding Clamp

PCNA and the β-sliding clamp associate with MutSα/β and MutS, respectively. Although initial reports suggested that the PCNA-MutSα complex may enhance mismatch recognition, it has been recently demonstrated that there is no apparent change in affinity of MutSα for a mismatch in the presence or absence of PCNA. Furthermore, mutants of MutSα that are unable to interact with PCNA *in vitro* exhibit the capacity to carry out mismatch recognition and mismatch excision to near wild type levels. Such mutants are defective in the repair reaction directed by a 5' strand break, suggesting for the first time MutSα function in a post-excision step of the reaction.

Clinical Significance

Inherited Defects in Mismatch Repair

Mutations in the human homologues of the Mut proteins affect genomic stability, which can result in microsatellite instability (MSI), implicated in some human cancers. In specific, the hereditary nonpolyposis colorectal cancers (HNPCC or Lynch syndrome) are attributed to damaging germline variants in the genes encoding the MutS and MutL homologues MSH2 and MLH1 respectively, which are thus classified as tumour suppressor genes. One subtype of HNPCC, the Muir-Torre Syndrome (MTS), is associated with skin tumors. If both inherited copies (alleles) of a MMR gene bear damaging genetic variants, this results in a very rare and severe condition: the mismatch repair cancer syndrome (or constitutional mismatch repair deficiency, CMMR-D), manifesting as multiple occurrences of tumors at an early age, often colon and brain tumors.

Epigenetic Silencing of Mismatch Repair Genes

Sporadic cancers with a DNA repair deficiency only rarely have a mutation in a DNA repair gene, but they instead tend to have epigenetic alterations that reduce DNA repair gene expression. About 13% of colorectal cancers are deficient in DNA mismatch repair, commonly due to loss of MLH1 (9.8%), or sometimes MSH2, MSH6 or PMS2 (all ≤1.5%). For most MLH1-deficient sporadic colorectal cancers, the deficiency was due to methylation of the MLH1 promoter. Other cancer types have higher frequencies of MLH1 loss (table below), which are again largely a result of methylation of the promoter of the *MLH1* gene. A different epigenetic mechanism underlying MMR deficiencies might involve over-expression of a microRNA, for example miR-155 levels inversely correlate with expression of MLH1 or MSH2 in colorectal cancer.

Cancers deficient in MLH1		
Cancer type	Frequency of deficiency in cancer	Frequency of deficiency in adjacent field defect
Stomach	32%	24%-28%
Stomach (foveolar type tumors)	74%	71%
Stomach in high-incidence Kashmir Valley	73%	20%
Esophageal	73%	27%
Head and neck squamous cell carcinoma (HNSCC)	31%-33%	20%-25%
Non-small cell lung cancer (NSCLC)	69%	72%
Colorectal	10%	

MMR Failures in Field Defects

A field defect (field cancerization) is an area of epithelium that has been preconditioned by epigenetic or genetic changes, predisposing it towards development of cancer. As pointed out by Rubin "there is evidence that more than 80% of the somatic mutations found in mutator phenotype human colorectal tumors occur before the onset of terminal clonal expansion." Similarly, Vogelstein et al. point out that more than half of somatic mutations identified in tumors occurred in a pre-neoplastic phase (in a field defect), during growth of apparently normal cells.

MLH1 deficiencies were common in the field defects (histologically normal tissues) surrounding tumors; (table above). Epigenetically silenced or mutated MLH1 would likely not confer a selective advantage upon a stem cell, however, it would cause increased mutation rates, and one or more of the mutated genes may provide the cell with a selective advantage. The deficient *MLH1* gene could then be carried along as a selectively near-neutral passenger (hitch-hiker) gene when the mutated stem cell generates an expanded clone. The continued presence of a clone with an epigenetically repressed *MLH1* would continue to generate further mutations, some of which could produce a tumor.

MMR Components in Humans

In humans, seven DNA mismatch repair (MMR) proteins (MLH1, MLH3, MSH2, MSH3, MSH6, PMS1 and PMS2) work coordinately in sequential steps to initiate repair of DNA mismatches. In addition, there are Exo1-dependent and Exo1-independent MMR subpathways.

Other gene products involved in mismatch repair (subsequent to initiation by MMR genes) in humans include DNA polymerase delta, PCNA, RPA, HMGB1, RFC and DNA ligase I, plus histone and chromatin modifying factors.

In certain circumstances, the MMR pathway may recruit an error-prone DNA polymerase eta (POLH). This happens in B-lymphocytes during somatic hypermutation, where POLH is used to introduce genetic variation into antibody genes. However, this error-prone MMR pathway may be triggered in other types of human cells upon exposure to genotoxins and indeed it is broadly active in various human cancers, causing mutations that bear a signature of POLH activity.

MMR and Mutation Frequency

Recognizing and repairing mismatches and indels is important for cells because failure to do so results in microsatellite instability (MSI) and an elevated spontaneous mutation rate (mutator phenotype). In comparison to other cancer types, MMR-deficient (MSI) cancer has a very high frequency of mutations, close to melanoma and lung cancer, cancer types caused by much exposure to UV radiation and mutagenic chemicals.

In addition to a very high mutation burden, MMR deficiencies result in an unusual distribution of somatic mutations across the human genome: This suggests that MMR preferentially protects the gene-rich, early-replicating euchromatic regions. In contrast, the gene-poor, late-replicating heterochromatic genome regions exhibit high mutation rates in many human tumors.

The histone modification H3K36me3, an epigenetic mark of active chromatin, has the ability to recruit the MSH2-MSH6 (hMutSα) complex. Consistently, regions of the human genome with high levels of H3K36me3 accumulate less mutations due to MMR activity.

Loss of Multiple DNA Repair Pathways in Tumors

Lack of MMR often occurs in coordination with loss of other DNA repair genes. For example, MMR genes *MLH1 and MLH3* as well as 11 other DNA repair genes (such as *MGMT* and many NER pathway genes) were significantly down-regulated in lower grade as well as in higher grade astrocytomas, in contrast to normal brain tissue. Moreover, MLH1 and MGMT expression was closely correlated in 135 specimens of gastric cancer and loss of MLH1 and MGMT appeared to be synchronously accelerated during tumor progression.

Deficient expression of multiple DNA repair genes is often found in cancers, and may contribute to the thousands of mutations usually found in cancers.

DNA Sequencing

DNA sequencing is a laboratory method used to determine the sequence of a DNA molecule. Sequencing DNA means determining the order of the four chemical building blocks – called "bases" – that make up the DNA molecule. The sequence tells scientists the kind of genetic information that is carried in a particular DNA segment. For example, scientists can use sequence information to determine which stretches of DNA contain genes and which stretches carry regulatory instructions, turning genes on or off. In addition, and importantly, sequence data can highlight changes in a gene that may cause disease. In the DNA double helix, the four chemical bases always bond with the same partner to form "base pairs." Adenine (A) always pairs with thymine (T); cytosine (C) always pairs with guanine (G). This pairing is the basis for the mechanism by which DNA molecules are copied when cells divide, and the pairing also underlies the methods by which most DNA sequencing experiments are done. The human genome contains about 3 billion base pairs that spell out the instructions for making and maintaining a human being. In 1980 Frederick Sanger was awarded the Nobel Prize in chemistry for his contributions to understanding DNA sequences.

In the early 1990s Sanger sequencing was used for sequencing a large fraction of genomes currently used in modern databases, including the human genome. In Sanger sequencing, the DNA to be sequenced serves as a template for DNA synthesis. A DNA primer is designed to be a starting point for DNA synthesis on the strand of DNA to be sequenced. Four individual DNA synthesis reactions are performed. The four reactions include normal A, G, C, and T deoxynucleotide triphosphates (dNTPs), and each contains a low level of one of four dideoxynucleotide triphosphates (ddNTPs): ddATP, ddGTP, ddCTP, or ddTTP. The four reactions can be named A, G, C and T, according to which of the four ddNTPs was included. When a ddNTP is incorporated into a chain of nucleotides, synthesis terminates. This is because the ddNTP molecule lacks a 3′ hydroxyl group, which is required to form a link with the next nucleotide in the chain.

Since the ddNTPs are randomly incorporated, synthesis terminates at many different positions for each reaction. Following synthesis, the products of the A, G, C, and T reactions are individually loaded into four lanes of a single gel and separated using gel electrophoresis, a method that separates DNA fragments by their sizes. The bands of the gel are detected, and then the sequence is read from the bottom of the gel to the top, including bands in all four lanes. For instance, if the lowest band across all four lanes appears in the A reaction lane, then the first nucleotide in the sequence is A. Then if the next band from bottom to top appears in the T lane, the second nucleotide in the sequence is T, and so on. Due to the use of dideoxynucleotides in the reactions, Sanger sequencing is also called "dideoxy" sequencing.

The modern rapid methods for DNA sequencing fall into two broad groups depending on the procedures used to generate and relate sets of labelled oligonucleotides, after resolution by gel electrophoresis, permit the DNA sequence to be deduced. The first group of methods employs a primed synthesis approach in which a single stranded template, containing or comprising the sequence of interest, is copied to produce the radioactively labelled complementary strand. Sets of partially elongated molecules are produced by using chain-termination inhibitors. These molecules can then be fractionated on denaturing gels. The patterns of labelled bands obtained will be used to deduce the sequence. This process is very adaptable and originally devised for sequencing naturally occurring single stranded DNAs. Highly efficient procedures have now been developed for generating single stranded templates from any duplex DNA. The forerunner of the primed synthesis method was the plus and minus method developed by Sanger and Coulson in 1975. By itself the method has distinct limitations and additional procedures need to be employed to confirm regions of the sequences deduce. One of these procedures is the depurination method, which was originally introduced by Burton and Peterson. Another development which made more flexible was the single site ribosubstitution reaction. Further developments in primed synthesis methods were made possible by the introduction of the chain terminating dideoxy-nucleotides as specific inhibitors.

DNA Sequencing by Primed Synthesis Methods

This method was developed by sanger et all in the year 1973. This method was used to determine two particular nucleotide sequences in bacteriophage DNA using DNA polymerase primed by synthetic oligonucleotides. The target oligonucleotide is hybridized to a specific complementary region on the single strand DNA. Deoxynucleotides is added sequentially by DNA polymerase to the 3'-OH end of the primer. Using [32P]-labelled deoxyribonucleoside 5'-triphosphates a radioactive complimentary copy of a defined region of the template is obtained. Next stage of development on this method came after two years. Analysis of the complementary DNA is done by fractionating the DNA fragments by electrophoresis on high resolution by polyacrylamide gels. It constitutes a relatively rapid and simple method for sequence analysis and illustrates the principle on which the modern chain-termination procedure is based.

Principle of the 'Plus and Minus' Method

Primed Synthesis

A DNA primer (commonly a restriction fragment or synthetic oligonucleotide) is hybridized to the single-stranded DNA template and the primer is extended to a limited degree with DNA polymerase I in the presence of all four deoxynucleoside 5'- triphosphates one of which is labelled. Samples of the reaction mixtures are taken at different time periods to find the degree of extension of primers. The reaction is terminated by addition of EDTA and the various samples are then combined. DNA polymerase is removed by phenol extraction. Extended polynucleotide chain, still hybridized to the template, is separated from the excess deoxynucleoside triphosphates by gel filtration on sephadex or agarose. The product is a mixture of partially elongated fragments with variable chain length. Ideally the product is a mixture of polynucleotides in which all possible chain lengths of the complementary strand are present corresponding to an elongation of the primer from 0 to 200 nucleotides.

The 'Minus' Reaction

The 'minus' reaction utilizes the same principle. The random mixture of oligonucleotides, still annealed to the DNA template is re-incubated with DNA polymerase in the presence of only three deoxynucleoside triphosphates. Synthesis then proceeds as far as the missing triphosphate on each chain. Four separate reactions are set up, each with one of the triphosphates missing. After incubation the reaction mixtures are denatured to separate the nascent strands from the template and the four samples simultaneously analyzed by electrophoresis on a high resolution polyacrylamide gel in the presence of 8M urea and the separated oligonucleotides visualized by autoradiography. On the acrylamide gel mobility is essentially proportional to size so that the various oligonucleotide products, all of which should have a common S-end will be arranged in a ladder according to size. Each oligonucleotide should be resolved from its neighbour, which contains one more residue, by a distinct space. As would be expected the resolution falls off with increasing chain length and it is this factor which usually determines how far the autoradiograph may be read. The separation between two fragments of say 150 and 151 nucleotides will be much less than that between two fragments of 20 and 21 nucleotides. The autoradiograph of the -A channel will consist of a set of bands, each of which corresponds to an extension product up to, but not including, the next A residue in the sequence. Thus the positions of the A residues are located. In a similar way the positions of the other residues are located from the sequencing channels and, in principle, the sequence of the DNA is read off from the autoradiograph. Usually, however, this system alone is not sufficient to establish the sequence and a second line of attack, the 'plus' system is used to confirm and complement the data from the 'minus' system.

The 'Plus' System

In the year 1971, 1972 Englund observed that in the presence of a single deoxynucleoside triphosphate, DNA polymerase from phage T4 infected E. coli (T4 polymerase) will degrade double-stranded DNA from its 3'-ends but this 3'-exonuclease activity will stop at residues corresponding to the single deoxynucleoside triphosphate present in the reaction mixture. Since T4 polymerase lacks the 5'-exonuclease activity found in E. coli polymerase incubation with this enzyme serves to trim back each elongated product to the residue corresponding to the added deoxynucleotide. The polymerizing activity of the enzyme catalyses the turnover of this residue but effectively halts the progress of the 3'-exonuclease activity. In the 'plus' reaction this method is applied to four further samples of the primer template complex isolated above. Samples are incubated with T4 polymerase and a single triphosphate and, after denaturation, the products are analysed directly by gel electrophoresis. Thus, in the +A system only dATP is added and all the chains will consequently terminate with a terminal A residue. The bands observed on the gel will therefore correspond to a set of fragments representing all the extension products which terminate with A. The products from the 'plus' reaction will be one residue longer than the corresponding band in the 'minus' A reaction.

The Polyacrylamide Gel: Interpretation of Results

The analysis of the products from the 'plus' and 'minus' reactions demands an acrylamide gel capable of resolving oligonucleotides which differ in length by only one residue. A 12% polyacrylamide slab gel in a Tris-borate buffer containing 8M-urea is usually employed for this purpose. Since the adequate resolution of products is the main limiting factor in this type of analysis and since this in turn depends on obtaining a sharp autoradiograph it is important to use extremely thin gels to avoid the blurring which would unavoidably result from the high energy p-emission of [PI embedded in a thicker gel. We define the smallest oligonucleotide in the -T channel as band 1. This means that the next residue after the 3'-terminus of the oligonucleotide corresponding to band 1 will be a T.

This is equivalent to, and is confirmed by the presence of a band in the +T channel corresponding to the next longest oligonucleotide. Band 2 occurs in the +T channel and -A channel showing that its 3'-terminus is T and the adjacent nucleotide is an A, thus defining the dinucleotide sequence T-A. The next longest oligonucleotide occurs in the +A and -C channels. This defines the dinucleotide A-C and so extends the sequence to T-A-C. Typically a sequence of 60-100 nucleotides, starting about 10-20 nucleotides from the 3'-end of the primer sequence, can be obtained from a single gel. In the above example each nucleotide in the sequence is represented by bands in both the + and − channels. However, if a run of two or more identical nucleotides occur, only the first one will be seen in the minus reaction and only the last one of the run will be seen in the equivalent plus reaction. If a synthetic oligonucleotide or a small restriction fragment (<loo nucleotides) is used as a primer for the initial extension, the products of the plus and minus reactions can be analysed directly.

Clearly, the smaller the primer that can be used, consistent with its ability to yield a unique primer-template complex in the annealing reaction, the greater the amount of sequence information that can be deduced since the extension reactions can be pushed further and still yield resolvable fragments. When using primers of this sort it is important to maintain the integrity of the 5'-terminus so that the difference in length of the fragments depends only on differences at their 3' termini. This is achieved by using DNA polymerase lacking the normal 5'- exonuclease activity of DNA polymerase I. If longer restriction fragments are used as primers it becomes necessary to cleave the primer from the can be deduced for the synthesized oligonucleotide. Typically a sequence of 60-100 nucleotides, starting about 10-20 nucleotides from the 3'-end of the primer sequence, can be obtained from a single gel. In the above example each nucleotide in the sequence is represented by bands in both the + and − channels. However, if a run of two or more identical nucleotides occur, only the first one will be seen in the minus reaction and only the last one of the run will be seen in the equivalent plus reaction.

The distance separating the different bands, in either the plus or minus patterns, will define how many residues there are in a run. This can lead to problems in the precise determination of the lengths of longer 'runs' and, partly for that reason, it is advantageous to include a 'zero' channel on the sequencing gel. This is simply an aliquot of the initial reaction which ideally will contain labelled oligonucleotides of all possible chain lengths and will therefore yield distinct bands corresponding to each residue of a run. This is conveniently done by digestion with the restriction enzyme originally used to prepare the primer. The products for analysis in this case represent only the denovo synthesized sequences and in consequence are theoretically capable of yielding relatively more sequence data than in experiments where the primer remains attached to the analysed products. Some restriction enzymes, however, are inhibited by the single-stranded DNA present in uncopied regions of the template and these enzymes cannot therefore be reliably used to cleave the primer from the extended product. One way round this problem is to use the single-site ribosubstitution method, in which a single ribonucleotide is incorporated at the priming site thus allowing the primer to be cleaved from the extended product with ribonuclease or alkali. This method is also useful if the restriction endonuclease used to generate the primer has a second cleavage site within the region to be sequenced.

Additional methods useful in conjunction with plus and minus method:

Ribosubstitution Method

This method was developed by Brown in the year 1978. In the protocol described above, restriction fragments are used as primers for the synthesis of cDNA, and the same endonuclease is subsequently used to remove the primer and generate a unique 5'-terminus on the cDNA. However, as pointed out earlier, some restriction enzymes are strongly inhibited by the single stranded regions present in the template and cannot be used to cleave at the restriction site. An additional problem arises if a second cleavage site for the same enzyme is present within the sequence copied into the radioactive DNA. Two sets of fragments would be generated and a unique sequence would not be obtained. In the ribosubstitution method these problems are circumvented by the addition of one or more ribonucleotides between the DNA primer and the radioactive cDNA. This site is susceptible to cleavage with ribonuclease or alkali. In the presence of Mn++ ions, E. coli DNA polymerase I will incorporate ribonucleotides into DNA. In the single site ribosubstitution reaction a ribonucleotide is incorporated at the 3'-end of a DNA primer in the presence of Mn2+, and with no other triphosphates present. Further ribonucleotide incorporation is effectively suppressed in the subsequent elongation reaction by the addition of deoxyribonucleoside triphosphates.

The plus and minus method indeed is a major breakthrough in the development of primed synthesis method for the rapid development of DNA sequences. However, there are certain limitations and one major limitation is that it cannot be applied directly to double stranded DNA. So a strand separation of either the template or primer must first be carried out. However with the advent of cloning techniques the problem of obtaining single stranded DNAs from naturally double stranded molecules has now been solved. As pointed by sanger et al in the year 1977 the plus and minus method cannot be regarded as a completely reliable method in the absence of confirmatory data. The additional methods ribosubstitution and depurination have been useful in overcoming two of the problems that arise.

The field of DNA sequencing has a diverse history. In 1970s and for almost another decade, DNA sequencing was barely automated and therefore very tedious process which allowed determining only a few hundred nucleotides in an experiment. In the late 1980s, semi-automated sequencers with higher throughput became available, still only able to determine a few sequences at a time. One breakthrough in the early 1990s was the development of capillary array electrophoresis and appropriate detection systems. Only within the last five years, alternative sequencing strategies like pyrosequencing, reversible terminator chemistry, sequencing-by-ligation, virtual terminator chemistry and real-time sequencing were developed or converged into new instruments. These new instruments require us to completely redefine the term "highthroughput sequencing", as they outperform the older Sanger-sequencing technologies by a factor of 100 to 1,000 in daily throughput and reduce the cost of sequencing one million nucleotides (1 Mb) to 4%-0.1% of that associated with Sanger sequencing. In the past few years a number of sequencing technologies have been developed that are parallelizable and therefore able to create more sequence output compared to conventional Sanger sequencing. These are collectively called next-generation sequencing (NGS) approaches.

Next Generation Sequencing

A DNA polymerase-based method was developed by Sanger and Coulson in the 1970s for DNA sequencing. It was the first enzyme-based approach and is known as Sanger sequencing. This method was used to sequence bacteriophage ΦX174 for the first time. From the 5' end up to 80 nucleotides were accurately identified and the following 50 nucleotides were identified with lower accuracy. In the following years enzyme based methods were optimized. The primary limitation was low throughput due to the template preparation and the enzymatic reactions. Sanger sequencing based on the automated sequencers was able to sequence up to 384 reads in parallel and about 1,000 bases per read per hour. The sequencing cost dropped rapidly due to advances in the sequencing methods.

Three major providers of next generation DNA sequencing are Illumina/Solexa, Roche/454 and Life/APG. Their methods include template preparation and sequencing followed by imaging and image data analysis. Specific changes in the protocols separate these methods. The two common approaches used during the template preparation step are either using clonally amplified templates or single molecule templates. Although these approaches differ in biochemistry they all follow the principle of cyclic-array sequencing. These technologies have been released as commercial products, e.g., the Solexa Genome Analyzer, the SOLiD platform, 454 Genome Sequencers, and the HeliScope Single Molecule Sequencer technology. These technologies create reads of length 25 – 250 bps and with up to 40 million reads per run.

Sanger Capillary Sequencing

Current Sanger capillary array electrophoresis (CAE) systems, are based on the same general scheme applied in 1977 for the φX174 genome: First, millions of copies of the sequence to be determined are purified or amplified, depending on the source of the sequence. Reverse strand synthesis is performed on these copies using a known priming sequence upstream of the sequence to be determined and a mixture of deoxy-nucleotides (dNTPs, the standard building blocks of DNA) and dideoxy-nucleotides (ddNTP, modi-fied nucleotides missing a hydroxyl group at the third carbon atom of the sugar). The dNTP/ddNTP mixture causes random, non-reversible termination of the extension reaction. Further denaturation and clean-up of free nucleotides, primers and the enzyme is performed. Then the resulting molecules are sorted by their molecular weight (corresponding to the point of termination). Then the label attached to the terminating ddNTPs is read out sequentially in the order created by the sorting step. Sorting by molecular weight was originally performed using gel electrophoresis but is nowadays carried out by capillary electrophoresis.

Originally, radioactive or optical labels were applied in four different terminator reactions. While today with the advent of more sensitive detection techniques four different fluorophores, one per nucleotide (A, C, G and T) are used in a single reaction. In the modern sequencing reactions several rounds of primer extensions (equivalent to a linear amplification) and more sensitive detection systems permit smaller amounts of starting DNA to be used for. Unfortunately, there is still little automation for creation of the high copy input DNA with known priming sites and it is done by cloning. In the cloning procedure the target sequence is introduced into a known vector sequence using restriction and ligation process. Then a bacterial strain is used to amplify the target sequence in vivo – thereby exploiting the low amplification error due to inherent proof-reading and repair mechanisms. However this process is very tedious. Later integrated microfluidic devices have been developed which aim to automate the DNA extraction, in vitro amplification and sequencing on the same chip. Using current Sanger sequencing technology, it is technically possible for up to 384 sequences of between 600 and 1,000 nucleotides (nt) in length to be sequenced in parallel. The sequencing error observed for Sanger sequencing is mainly due to errors in the amplification.

454/Roche Genome Sequencer

In contrast to the Sanger technology, pyrosequencing is based on iteratively complementing single strands and simultaneously reading out the signal emitted from the nucleotide being incorporated (also called "sequencing by synthesis" or "sequencing during extension"). As the read out is done simultaneously with the sequence extension so electrophoresis is no longer required to generate an ordered read out of the nucleotides. In the pyrosequencing process, one nucleotide at a time is washed over several copies of the sequence to be determined, causing polymerases to incorporate the nucleotide if it is complementary to the template strand. The incorporation stops if the longest possible stretch of complementary nucleotides has been synthesized by the polymerase.

In the process of incorporation, one pyrophosphate per nucleotide is released and converted to adenosine triphosphate (ATP) by an ATP sulfurylase. The ATP drives the light reaction of luciferases present and the emitted light signal is measured. To prevent the deoxyadenosine triphosphate (dATP) provided in a typical sequencing reaction from being used directly in the light reaction, deoxy-adenosine-5'-(alpha-thio)-triphosphate (dATPαS), which is not a substrate of the luciferase, is used for the base incorporation reaction of adenine. After capturing the light intensity, the remaining unincorporated nucleotides are washed away and the next nucleotide is provided. Free nucleotides are then washed over the sequencing plate and the light emitted during the incorporation is captured for all wells in parallel using a high resolution CCD camera, exploiting the light-transporting features of the plate used. One of the main prerequisites for applying array-based pyrosequencing approach is covering individual beads with multiple copies of the same molecule. This is done by first creating sequencing libraries in which every individual molecule gets two different adapter sequences, one at the 5' end and one at the 3' end of the molecule. In the case of the 454/Roche sequencing library preparation, this is done by sequential ligation of two pre-synthesized oligos. One of the adapters added is complementary to oligonucleotides on the sequencing beads and thus allows molecules to be bound to the beads by hybridization.

Low molecule-to-bead ratios and amplification from the hybridized double-stranded sequence on the beads (kept separate using polymerase chain reaction in an water-in-oil emulsion, i.e emulsion PCR) makes it possible to grow beads with thousands of bound copies of a single starting molecule. Using the second adapter, beads covered with molecules can be separated from empty beads (using capture beads with oligonucleotides complementary to the second adapter) and are then used in the sequencing reaction as described above.

The average substitution error rate is in the range of 10–3 to 10–4, which is higher than the Sanger sequencing but is the lowest among new sequencing technologies As mentioned earlier for Sanger sequencing, in vitro amplifications performed for the sequencing preparation cause a higher background error rate, that is, the error introduced into the sample before it enters the sequencing process. In addition, in bead preparation (i.e. emulsion PCR step) a fraction of the beads end up carrying copies of multiple different sequences. These "mixed beads" will participate in a high number of incorporations per flow cycle, resulting in sequencing reads that do not reflect real molecules. Most of these reads are automatically filtered during the software post-processing of the data. The filtering of mixed beads may however cause a depletion of real sequences with a high fraction of incorporations per flow cycle. Most of these problems can be resolved by higher coverage. Strong light signals in one well of the picotiter plate may also result in insertions in sequences in neighbouring wells. If the neighbouring well is empty this can generate so-called ghost wells, i.e. wells for which a signal is recorded even though they contain no sequence template, hence the intensities measured are completely caused by bleed-over signal from the neighbouring wells. Error rate in the case of 454 sequencing, is caused by (1) a reduction in enzyme efficiency, (2) some molecules on the beads no longer being elongated and (3) by an increasing, so-called, phasing effect. Phasing is observed when a population of DNA molecules amplified from the same starting molecule (ensemble) is sequenced, and describes the process whereby not all molecules in the ensemble are extended in every cycle. This causes the molecules in the ensemble to lose synchrony/phase, and results in an echo of the preceding cycles to be added to the signal as noise. The current 454/Roche GS FLX Titanium platform makes it possible to sequence about 1.5 million such beads in a single experiment and to determine sequences of length between 300- 500nt. This is largely due to limitations imposed by the efficiency of polymerases and luciferases which drops over the sequencing run resulting in decreased base qualities.

Illumina Genome Analyzer

The reversible terminator technology used by the Illumina Genome Analyzer employs the sequencing similar to sangers sequencing concept. The incorporation reaction is stopped after each base, the label of the base incorporated is read out with fluorescent dyes and the sequencing reaction is then continued with the incorporation of the next base. Similar to 454/Roche, the Illumina sequencing protocol also requires that the sequences to be determined are converted into a sequencing library, which allows them to be amplified and immobilized for sequencing. For this purpose two different adapters are added to the 5' and 3' ends of all molecules using ligation of so-called forked adapters2.

The library is then amplified using longer primer sequences which extend and further diversify the adapters, i.e. add further unique nucleotides at both adapter ends, to create the final sequence needed in subsequent steps. This double-stranded library is melted using sodium hydroxide to obtain single stranded DNAs, which are then pumped at a very low concentration through the channels of a flow cell. This flow cell has on its surface two populations of immobilized oligonucleotides complementary to the two different single stranded adapter ends of the sequencing library. These oligonucleotides will hybridize to the single stranded library molecules.

By reverse strand synthesis starting from the hybridized (double-stranded) part, the new strand being created is covalently bound to the flow cell. If this new strand bends over and attaches to another oligonucleotide complementary to the second adapter sequence on the free end of the strand, it can be used to synthesize a second covalently bound reverse strand. This process of bending and reverse strand synthesis, called bridge amplification, is repeated several times and creates what are termed clusters, the accumulation of several thousand copies of the original sequence in very close proximity to each other on the flow cell. These randomly distributed clusters contain molecules that represent the forward as well as reverse strands of the original sequences. Before determining the sequence, one of the strands has to be removed to prevent it from hindering the extension reaction sterically or by complementary base pairing. Selective strand removal targets base modifications of the oligonucleotide populations on the flow cell. Following strand removal, each cluster on the flow cell consists of single stranded, identically oriented copies of the same sequence; which can be sequenced by hybridizing the sequencing primer onto the adapter sequences and starting the reversible terminator chemistry. "Solexa sequencing", as it was introduced in early 2007, initially allowed for the simultaneous sequencing of several million very short sequences (at most 26nt) in a single experiment. In recent years the Illumina Genome Analyzer (GA), has increased flow cell cluster densities (more than 300 million clusters per run), a wider range of the flow cell is imaged, and sequence reads of up to 125nt can be generated. This is achieved by chemical melting and washing away the synthesized sequence, repeating a few bridge amplification cycles for reverse strand synthesis and then selectively removing the starting strand (again using base modifications of the flow cell oligonucleotide populations), before blocking 3' ends and annealing another sequencing primer for the second read. Using this "paired end sequencing" approach, approximately twice the amount of data can be generated.

The Illumina library and flow cell preparation includes several in vitro amplification steps which cause a high background error rate and contribute to the average error rate of about $10-2$ to $10-3$. Further, the flow cell preparation creates a fraction of ordinary looking clusters which are initiated from more than one individual sequence. These result in mixed signals and mostly low quality sequences for these clusters. In an effect similar to the 454 ghost wells, the Illumina image analysis software may identify reagent crystals, dust and lint particles as clusters and call sequences from these. Similar to the other platforms, the error rate increases with increasing position in the determined sequence. This is mainly due to phasing, which increases the background noise as sequencing progresses. While the ensemble sequencing process for pyrosequencing creates unidirectional phasing from lagging, non-extended molecules, reversible terminator sequencing creates bi-directional phasing as some incorporated nucleotides may also fail to be correctly terminated – allowing the extension of the sequence by another nucleotide in the same cycle.

The Genome Analyzer uses four fluorescent dyes to distinguish the four nucleotides A, C, G and T. Of these, two pairs (A/C and G/T) excited using the same laser, are similar in their emission spectra and show only limited separation using optical filters. Therefore, the highest substitution errors observed are between A/C and G/T. Even though the Illumina Genome Analyzer reads show a higher average error rate and are considerably shorter than 454/Roche reads, this instrument determines more than 10,000 Mb per day with a price of about 0.50$/Mb. This is more than ten times higher daily throughput than 454/Roche and for a considerably lower price per megabase.

Helicos HeliScope

This sequencer is able to sequence individual molecules instead of molecule ensembles created by an amplification process. The advantage with single-molecule sequencing is that it is not affected by biases or errors introduced in a library preparation or amplification step. One more advantage is it may facilitate sequencing of minimal amounts of input DNA, which is requisite for many research projects. Using methods able to detect non-standard nucleotides, it could also allow for the identification of DNA modifications, commonly lost in the in vitro amplification process.

The HeliScope instrument was installed in few numbers which might be surprising given the advantages of single molecule sequencing. This also reflects the specific limitations of this platform, the price (about one million dollars), and a relatively small market. The technology applied by the HeliScope could be termed asynchronous virtual terminator chemistry. Input DNA is fragmented and melted before a poly-A-tail is synthesized onto each single stranded molecule using a polyadenylate polymerase. In the last step of polyadenylation, a fluorescently labelled adenine is added. The library, i.e. the polyadenylated single stranded DNA, is washed over a flow cell where the poly-A tails bind to poly-T-oligonucleotides. The bound coordinates on the flow cell are determined using a fluorescence-based read out of the flow cell. Having these coordinates identified, the fluorescent label of the 3' adenine is removed and the sequencing reaction started. Polymerases are washed through the flow cell with one type of fluorescently labelled nucleotides (A, C, G and T) at a time and the polymerases extend the reverse strand of the sequences starting from the poly-T-oligonucleotides. The nucleotide incorporation of the polymerases is slowed down by the fluorescent labelling and allows for at most one incorporation before the polymerase is washed away together with the non-incorporated nucleotides.

The flow cell is then imaged again, the fluorescent dyes are removed and the reaction continued with another nucleotide. By this process not every molecule is extended in every cycle, which is why it is an asynchronous sequencing process resulting in sequences of different length. Since single molecules are sequenced, the signals being measured are weak. Also there is no possibility that mis incorporation errors can be corrected by an ensemble effect. Due to the fact that molecules are attached to the flow cell by hybridization only, there is a chance that template molecules can be lost in the wash steps. In addition, molecules may be irreversibly terminated by the incorporation of incorrectly synthesized nucleotides. Overall, reads which are 24 to 70nt are shorter than the other platforms.

Single Molecule Real Time (SMRT)

Another technology for sequencing individual molecules is SMRT (Single Molecule Real Time) sequencing technology, which belongs to Pacific Biosciences. This technology performs the sequencing reaction on silicon dioxide chips. Chips are of 100nm metal film containing thousands of tens-of nanometers diameter holes, so called zero-mode waveguides (ZMWs). Each ZMW is used as a nano visualization chamber, providing a detection volume of about 20 zeptoliters (10−21 liters). At this volume, a single molecule can be illuminated while excluding other labeled nucleotides in the background – saving time and sequencing chemistry by omitting wash steps. For SMRT sequencing, a single DNA polymerase is fixed to the bottom of the surface within the detection volume of each ZMW. Nucleotides with different fluorescent dyes attached to the phosphate chain are used in concentrations allowing normal enzyme processivity. As the polymerase incorporates complementary nucleotides, the nucleotide is held within the detection volume for tens of milliseconds, orders of magnitude longer than for unspecific diffusion events. This way the fluorescent dye of the incorporated nucleotide can be identified during normal speed reverse strand synthesis. Further, by attaching the fluorescent dyes to the phosphate chain of the deoxy-nucleotides the dye is released with the cleaved pyrophosphate, generating an unmodified complimentary DNA strand. In pilot experiments, Pacific Biosciences showed that their technology allows for direct sequencing of a few thousand bases before the polymerase is denatured due to optic and thermal stress from the laser read-out of the dyes. They were also able to show that they can measure differences in polymerase kinetics to such an extend that modified nucleotides may be detected.

Repetitive DNA

Repetitive DNA is generally confined to higher eucaryotes. Some 20 to 40% of mammalian genomes consist of either highly repeated (more than 10 s copies per haploid genome) or moderately repeated (about 10^3 to lo^5 copies per haploid genome) sequences. By contrast, lower eucaryotes and procaryotes have smaller genomes, consisting of mostly low or single copy number coding sequences only.

The existance of repeated sequences in higher eucaryotes was generally established by one of three methods : A fraction of the genome reannealed very rapidly after denaturation (this being multiple copy number sequences, whereas single copy sequences reannealed much slower); secondly, centrifugation of genomic DNA in cesium chloride (CsCl) gradients (this method separated fractions of DNA that were of different buoyancy to the main band DNA and these so-called "satellites" were found to be highly repeated sequences); and thirdly, restriction of whole genomic DNA by some enzymes (for example, Hae III), separation of the fragments by electrophoresis in agarose gels, and staining with ethidium bromide (this showed the presence of distinct banding resulting from repeated sequences upon a background smear of DNA fragments).

A system of nomenclature and classification of repetitive DNA has evolved, this being largely based on the structural organization and reiteration frequency of each species. There are two broad classes, namely (1) the tandemly repetitive sequences and (2) the interspersed repetitive sequences. These two classes will be discussed.

Tandem Repetitive Sequences

Such sequences comprise some 5 to 10% of mammalian genomes and are characterized by the head-to-tail tandem repetition of lengths of DNA of generally some common sequence.

Table: Classtfication of repentive DNA in human genomes.

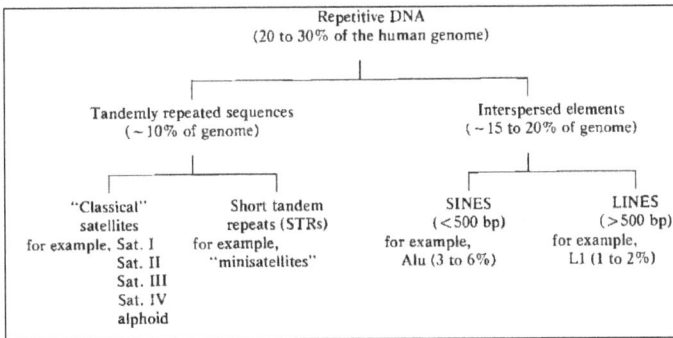

		Repetitive DNA (20 to 30% of the human genome)		
Tandemly repeated sequences (~ 10% of genome)			Interspersed elements (~ 15 to 20% of genome)	
"Classical" satellites for example, Sat. I Sat. II Sat. III Sat. IV alphoid	Short tandem repeats (STRs) for example, "minisatellites"		SINES (<500 bp) for example, Alu (3 to 6%)	LINES (>500 bp) for example, L1 (1 to 2%)

There are four classical classes of tandem repeat sequences in the human genome: Satellite I (0.2 to 0.5%), Satellite II (1 to 2%), Satellite III (1 to 3%), and Satellite IV (0.5 to 2%), totalling about 5% of the genome. These classes are representative of the DNA "satellite" bands separated by CsCI centrifugation of whole genomic DNA. As a consequence, the term "satellite" has become generally adopted to describe all tandemly repeated sequences (whether separable by centrifugation or not). This has lead further to the description of short tandem repeats (STRs) as "minisatellites" and recently longer tandem repeats as "midisatellites".

A further classical repetitive DNA is the alphoid class. This was originally found in abundance in higher primates, but is also present in human genomes (1 to 2%). A large proportion of human satellite DNA (particularly alphoid and Satellite I) is located at centromeres of chromosomes (centromeric repetitive DNA).

The simplest tandem structure and base sequences for each of the major classical human satellite DNAs are known. The lengths of the simplest repeating unit in each class is generally constant, but sequence divergence within these units is possible, thus generating a "family" of sequences within each class. The repeat units may be as small as 5 base pairs (bp) (for example, Satellite III and Satellite II being structurally related), but more typically are 170 to 250 bp in length. Often these can be demonstrated as "ladders" of DNA lengths on electrophoretic separations, the "rungs" being multiples of the basic repeat unit either as a result of the loss of the appropriate restriction site at the junction of individual units or as a result of incomplete digestion.

The lengths of tandem repeat sequences revealed experimentally depends entirely upon the restriction enzyme used. For example, the long arm of the Y chromosome contains about 5000 copies of a 3.4 kilobase (kb) tandem repeat (for example, Hae III digestion) accounting for as much as 40% of the whole chromosome. Many of these 3.4-kb repeats contain short tandem repeats constructed as pentameric (5-bp) units of a moderately divergent base sequence.

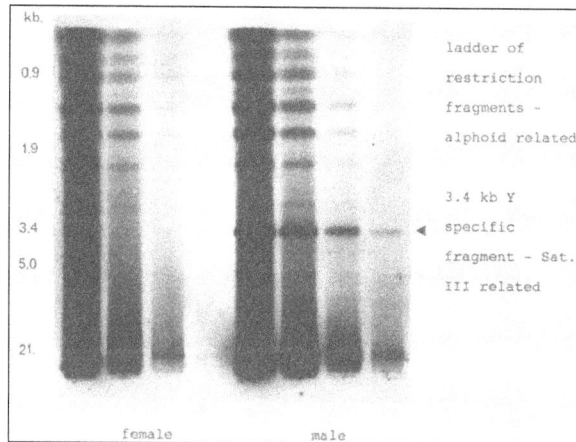

Demonstration of a "'ladder" of restriction fragments (ustng an alphoid sequence related probe) and a Y chromosome speciJ~c fragment (using a Satellite 111 sequence related probe). The DNA samples were decreastng concentrations of Hae 111 digested female and male DNA.

Thus each "classical" satellite generally consists of a diverged "family" of sequences, with particular restriction fragment lengths arising from specific chromosomes and the sequences of individual family members (in any one class) being chromosome specific (or nearly specific) in origin, for example, alphoid and Satellite III. This might have functional significance. For example, centromeric repetitive DNA may assist in chromosome recognition at meiotic divisions. Generally, however, the evolution and function of these "classical" satellites is still the subject of research.

Besides the "classical" satellites in human genomes, there are other tandemly repeated sequences. Some of these have been found to exist as a small single block of a tandemly repeated DNA sequence, at unique genomic loci and within noncoding DNA regions. Such blocks (often containing a common "core" portion to their sequences) may be dispersed at multiple genomic loci. The single most important property of these tandem repeats, first described in 1980, was their genonfic variability.

Polymorphism in Tandemly Repeated DNA

All polymorphisms associated with tandemly repeated DNA arise from restriction enzymes creating DNA fragment lengths that contain variable numbers of the repeated species. There are, in general, two contrasting causes by which this occurs. The distinction between these mechanisms is vital. In one case, the enzyme cuts the DNA externally to the block of tandem repeats, and in the other, the DNA is cut within the lengths of tandemly repeating units.

The consequence of this is that in the first mechanism any enzyme that cuts the DNA in regions flanking (but not within) the tandem repeat will reveal any variation that may exist in the number of repeats, and hence the fragment size. The second mechanism is enzyme specific, polymorphism being generally a result of an enzyme that cuts (or fails to cut) infrequently into the higher order repeat units of each tandem array.

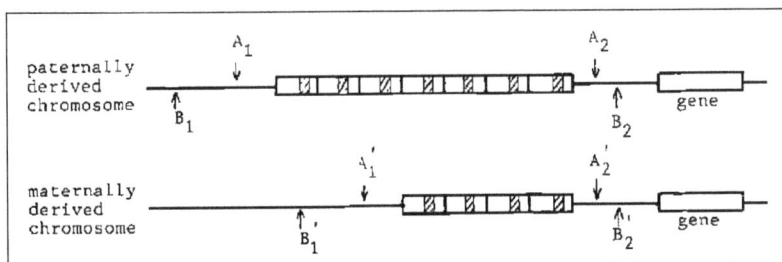

Schematic representation of a "minisatellite'" polymorphism at a smgle genomie locus. (Many "mh6satellites" have been Jbund near or within genes often because the gene and tts surroundings were being studied. Probing with a sequence containing both the "core" [hatched] and other site specific sequences [the blocked but nonhatched regions in the tandem repeats and solid lines immediately external to this] will generally show two restriction fragment lengths from a single locus: either A I A2 and A [A~ fragments using enzyme A or Bj B2 and B/B2 using enzyme B. Use of a "pol.veore" probe being tandem repeats of the hatched region only] may give complex patterns arising from more than one locus.

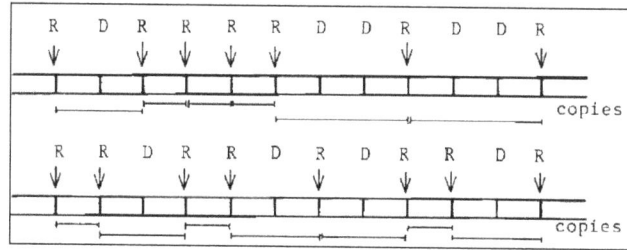

Schematic representation of a polymorphtsm m a long tandemly repeating satellite. The target sequence of the probe may be formed by all or part of the sequence in each repeat unit. Restriction sites may exist [R] or be deleted [D], giving rise to fragments of variable lengths. Some of these may be duplicated during evolution, giving multiple copies of the same length. Highly complex fragment patterns can arise from multiple loci of the same sequence.

Polymorphisms arising by the first mechanism have been referred to as "minisatellites", hypervariable regions (HVRs), variable number of tandem repeats (VNTRs), and variable length polymorphisms (VLPs). Polymorphisms arising by the second mechanism have also been referred to as VNTRs. As long as the reader remains conscious of the presumptive cause for each polymorphism, there appears to be no strong reason to prefer one name to another.

"Minisatellite" Polyrnorphism

The polymorphism in all such examples is an insertion/deletion mutational event, thereby lengthening or shortening the overall fragment length. The length of repeat units inserted or deleted is typically between 64 and 10 bp long. Tandem arrays of such units may exist at either a unique or a number of dispersed genomic sites. The supplementary references are given for extended information regarding each example.

The cause of a loss or gain in the number of repeat units at any genomic site may be either (1) integral numbers of unit slippage at replication or (2) unequal recombination between the tandemly repeated sequences or both. This recombination may indeed be directed by a particular GC rich "core sequence" which (with some sequence diversity) has been found to be embedded in many of the GC rich "minisatellites". This core is represented as the hatched regions within the repeats drawn in figure. Jeffrey's conclusion is that such core sequences, 11 to 16 bp in length in his examples, either promote the initial duplication of DNA immediately adjacent to it or assist in changing the number of repeat units so created by unequal recombination or faulty replication or both.

Supportive evidence is that the "core" sequence bears some sequence homology to the so-called "Chi" (crossover hotspot instigator) sequence believed to be active in procaryotes. The "Chi" sequence is postulated to be a target for a protein assisted recombination event between homologous DNA sequences, perhaps initiating alteration in the number of repeats and providing the interindividual variation. The "Chi" sequence has also been noted as showing homology to the X gene region of the hepatitis B virus, a sequence potentially recombinogenic as it is believed to be responsible for viral integration into human genomes.

The practical consequence of these details is as follows. If a "polycore" probe is hybridized under moderate stringency to suitably restricted DNA, the multiple loci (between 40 to 60) of this "core" sequence are revealed as complex patterns of restriction fragments (genetic "fingerprints"). This method of individualization has been applied to forensic science samples.

Table: STR (short tandem repeat) polymorphisms in the human genome mlnisatellites.

2a			
Source of Probe GC Rich Sequences	Variation in RFs "Alleles" (Enzyme)	Ref	Supp.
5' to insulin gene	21 (Sac I)	50	51-54
3' to H Ras 1onco gene	18 (Taq I)	55	56-59
"DI4SI" regions	>40 (Taq I)	55	58-61
Intron zeta 1 globin gene	4 (Hinf I)	62	63
lntragenic globin genes	3 (Pvu II)	62	64-66
3' to alpha globin gene	>20 (F'vu II)	64	49,66,67
"Fragment g,b	> 77 (Hinf I)	68	
lntron myoglobin genes	many (Hinf I)	27	47,69-76
Protein III gene M13 phage	many (Hae III)	77	78, 79
Clones hybridizing with synthetic oligomers			
for example, MLJ 14	>20 (Rsa I)	48	28,80,81
for example, YNH 24	> 20 (Msp I)	48	
lntron apolipoprotein C-II gene	many (EcoR I)	82	
2b			
Source of Probe-- "AT" Rich Sequences			
3' apolipoprotein B gene	>3 (Msp I	83	
3' type I1 collagen gene	3 (EcoR I)	84	
Pseudo-autosomal X\Y chromosome: for example, DXYS15	> 8 (Taq I)	85	

[a]Probe to multiple loci at reduced stringency of hybridization.

[b]Probe to multiple loci in human genomes.

If the probe, however, contains tandem repeats of both the "core" sequence and also flanking DNA specific to particular loci (represented by the nonhatched but blocked regions of each repeat and the solid lines external to these figure), then under high stringency conditions such a probe will generally detect only two (heterozygous) fragments from a single genomic locus. Probes of this sort have also been used in the analysis of forensic science samples If a "mixture" of single-locus probes is used simultaneously, a combined fragment pattern arises, while some single locus probes used at lower stringency conditions may reveal weaker hybridization from other loci, There are estimated to be at least 1500 "minisatellites" in the human genome. These include both GC and AT rich "cores" in the sequence repeats, suggesting that there may be different classes of sequences which may serve as primers to "minisatellite" formation. The AT rich "minisatellites" show no sequence homology with the "Chi" sequence, and other explanations for "minisatellite" formation and propagation are proposed.

This table gives an indication of the reported variability detected at each "minisatellite" loci. Extensive population frequency data and evidence of ethic variation exist for the variable length polymorphisms associated with the 3' region of the HRAS-I onco-gene and the "DI4SI" region. Other loci are yet to be as well characterized.

Some "minisatellite" core sequences show remarkable conservation throughout nature, the most interesting example thus far being a "minisatellite" found in the protein III gene of the "wild"-Type M 13 phage (a bacterial virus) used to locate VNTRs in human, bovine, equine, murine, and canine genomes.

Polymorphism in Long Tandem Repeats

Relatively complex human polymorphisms exist in sequences of long tandem repeats (LTRs); known examples have been listed in, including those found in the "classical" satellites, alphoid, and Satellite III. The distinction between such examples and those of the "minisatellites" is that polymorphism appears to be the result of either point mutational or more complex insertion/deletion/crossover events within the long tandem repeats. They are most likely to be detected using restriction enzymes which cut infrequently (or as a result of a site deletion fail to cut) into the higher order repeat units of each tandem array.

Generally, the polymorphisms in the LTR examples are seen experimentally as variable fragment lengths within more complex patterns, some or most of the fragments being constant between different individuals (There is also some quantitative (copy number) variation between individuals.

In the case of Satellite III, the probe used to establish the polymorphism has dual properties of hybridizing with a 3.4-kb Y specific repeat as well as hybridizing with a complex series of restriction fragments upon Taq 1 digestion of human genomes. The polymorphism may be the result of a specific base mutation in the small pentameric repeat typical of Satellite III.

Polymorphisms of the alphoid, Satellite III, or Bkm type (which contain multiple numbers of particular restriction fragment lengths) might be expected to be detected in smaller quantities of DNA than "minisatellite" polymorphisms (the fragments for which occur in only single copy per haploid genome, at either one or multiply dispersed loci). However, the "minisatellite" polymorphisms may be more discriminating between individuals.

Probes for LTRs (or indeed for interspersed repeats) that show no human polymorphisms will have an important role in forensic science analyses. Such probes will not only assist estimation as to the quantity of DNA under examination, and importantly the efficiency of its restriction, but will also demonstrate the extent of its degradation. The resultant invariant fragment patterns will be necessary as controls to probes producing polymorphic patterns. An advantage of DNA methodology is that this control information can be obtained by stripping and reprobing membranes with different probes.

Interspersed Repetitive Sequences

The second major class of repetitive DNA, comprising as much as 20% of the human genome, is composed of interspersed repetitive sequences These single units are scattered individually throughout the genome. They have been classified into two class sizes: short interspersed elements (SINES, less than 500 bp) and long interspersed elements (LINES, more than 500 bp).

Table: LTR (long eandem repeat) polymorphisms in the human genome.

Source of Probe for Polymorphism	Variation in RFs "Alleles" (Enzyme)	Ref	Supp.
Alphoid--chromosome X	4 (Hind lII)	90	91
--chromosome 17	3 (EcoR I)	90	
--chromosome I 1	4 (Xba I)	92	
--chromosome 6			
predominantly 5	(Taq I)	93	94-96
Satellite III	> 15 (Taq I)	97	98
Bkm	> 15 (Mbo I)	99	100-103
Sau3a	>6 (EcoR I)	104	105-107

The most abundant SINES in the human genome are the Alu repeats. These SINES amount to about 3 to 6% of the human genome or about 3×10^5 copies per haploid genome. On average they are spaced every 8 kb of the human genome, though there is evidence for their clustering. The Alu element (about 300 bp) is strongly sequence conserved and is bounded by short direct repeats. SINES of moderate homology to Alu are found in higher primates and rodents.

The most abundant LINES in human genomes are the L1 (Kpn) element. These elements are very variable in length, ranging from 7 kb down to about 500 bp, truncation occurring from the 5' end of the LI element towards the 3' end. The sequences occur about 1 to 4×10^4 times, constituting 1 to 2% of the human genome. Homologous L1 sequences are found in other mammals, particularly rodents.

The mechanism by which SINES and LINES are dispersed in multiple copies throughout the genome is not completely understood. However, both appear to be examples of mobile genetic elements (transposons). As a consequence of their origins (Alu likely being derived from a small RNA gene and Kpn perhaps from retroviral DNA), both have the ability to be transcribed (into RNA). The copied sequence may then be integrated (as DNA) into another genomic site, in a manner analogous to integration of retroviruses into eucaryotic genomes (termed retrotransposition).

The structural distinction made between tandem and interspersed repetitive sequences in should not be regarded as rigid, but rather extremes. The Sau 3a family of repeats (closely related to the alphoid family) consists of five tandem subunits (about 170 bp each), thus producing a "LINE-like" element (about 850 bp long). This can either exist independently or as tandem arrays of the 850-bp unit. Significantly, a small fraction of the 850-bp unit in the human genome is extrachromosomal, suggesting it to have properties of a mobile genetic element. The Sau 3a family of repeats shows moderate variable length polymorphism in human genomes.

Polymorphism in Interspersed Repeats

Interspersed repeats have thus far not disclosed the same degree of RFLPs as that associated with tandem repeats. One example is noteworthy. This is the L2(H) (Kpn 0.6 kb) human polymorphism, which perhaps is a result of unspecified sections of the human genome containing clusters (but not tandem repeats) of a particular Kpn core sequence. The arrangements of such sequences show both constant fragments as well as qualitative and quantitative variation of other fragments in the human genome.

Various Classes of Human Repetitive DNA Sequences

Telomere Repeat: The tandemly repeating unit TTAGGG located at the very ends of the linear DNA molecules in human and vertebrate chromosomes. The telomere repeat (TTAGGG)" extends for 5000 to 1.2,000 base pairs and has a structure different from that of normal DNA. A special enzyme called telomerase replicates the ends of the chromosomes in an unusual fashion that prevents the chromosome from shortening during replication.

Subtelomeric repeats: Classes of repetitive sequences that are interspersed in the last 500,000 bases of nonrepetitive DNA located adjacent to the telomere. Some sequences are chromosome specific and others seem to be present near the ends of all human chromosomes.

Microsatellite repeats: A variety of simple di-, tri-, tetra-, and penta-nucleotide tandem repeats that are dispersed in the euchromatic arms of most chromosomes. The dinucleotide repeat (GT)n is the most common of these dispersed repeats, occurring on average every 30,000 bases in the human genome, for a total copy number of 100,000. The GT repeats range in size from about 20 to 60 base pairs and appear in most eukaryotic genomes.

Minisatellite repeats: A class of dispersed tandem repeats in which the repeating unit is 30 to 35 base pairs in length and has a variable sequence but contains a core sequence 10 to 15 base pairs in length. Minisatellite repeats range in size from 200 base pairs up to several thousand base pairs, have lower copy numbers than microsatellite repeats, and tend to occur in greater numbers toward the telomeric ends of chromosomes.

Alu repeats: The most abundant interspersed repeat in the human genome. The Alu sequence is 300 base pairs long and occurs on average once every 3300 base pairs in the human genome, for a total copy number of 1 million. Alus are more abundant in the light bands than in the dark bands of giemsa-stained metaphase chromosomes. They occur throughout the primate family and are homologous to and thought to be descended from a small, abundant RNA gene that codes for the 300-nucleotide-long RNA molecule known as 7SL. The 7SL RNA combines with six proteins to form a protein-RNA complex that recognizes the signal sequences of newly synthesized proteins and aids in their translocation through the membranes of the endoplasmic reticulum (where they are formed) to their ultimate destination in the cell.

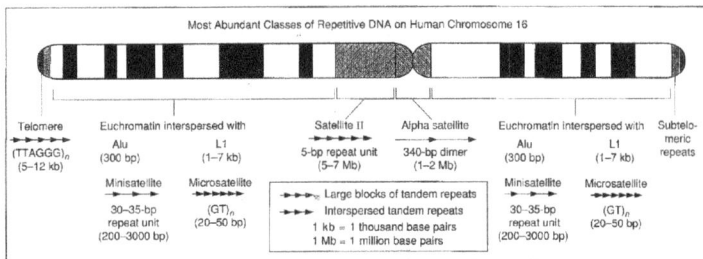

Most Abundant Classes of Repetitive DNA on Human Chromosome 16

LI repeats: A long interspersed repeat whose sequence is 1000 to 7000 base pairs long. L1s have a common sequence at the 3' end but are variably shortened at the 5' end and thus have a large range of sizes. They occur on average every 28,000 base pairs in the human genome, for a total copy number of about 100,000, and are more ablundant in Giemsa-stained dark bands. L1 repeats are also found in most other mammalian species. Full-length Lls (3.5 percent of the total) are a divergent group of class Ii retrotransposons—"ju roping genes" that can move around the genome and are thought to be remnants of retroviruses. [Class II retrotransposons have at least one protein-coding gene and contain a poly A tail (or series of As at the 3' end) as do messenger RNAs.] Recently, a full-length, functional LI was discovered. It was found to code for a functional reverse transcriptase-an enzyme essential to the process by which the Lls are copied and re-inserted into the genome.

Alpha satellite DNA: A family of related repeats that occur as long tandem arrays at the centromeric region of all human chromosomes. The repeat unit is about 340 base pairs and is a dimer, that is, it consists of two subunits, each about 170 base pairs long. Alpha satellite DNA occurs on both sides of the centromeric constriction and extends over a region 1000 to 5000 base pairs long. Alpha satellite DNA in other primates is similar to that in humans.

Satellite 1, H, and III repeats: Three classical human satellite DNAs, which can be isolated from the bulk of genomic DNA by centrifugation in buoyant density gradients because their densities differ from the densities of other DNA sequences. Satellite I is rich in As and Ts and is composed of alternating arrays of a 17- and 25-base-pair repeating unit. Satellites II and III are both derived from the simple five-base repeating unit AT-TCC, Satellite II is more highly diverged from the basic repeating unit than Satellite Ill. Satellites 1, II and III occur as long tandem arrays in the heterochromatic regions of chromosomes 1, 9, 16, 17, and Y and the satellite regions on the short (p) arms of chromosomes 13, 14, 15, 21, and 22.

Cotl DNA: The fraction of repetitive DNA that is separable from other genomic DNA because of its faster re-annealing, sequences that have copy numbers Number 20 1992 Los Akunos Science or renaturation, kinetics. Cot 1 DNA contains of 10,000 or greater.

References

- Polymerase-chain-reaction, science: britannica.com, Retrieved 27 February, 2019

- Polymerase-chain-reaction: abmgood.com, Retrieved 9 May, 2019

- Nucleic-acid-hybridization, recombinant-dna-technology: biotechnologynotes.com, Retrieved 2 March, 2019

- Dna-sequencing: mybiosource.com, Retrieved 18 July, 2019

Genetic Variation

In any particular population, the difference in DNA sequence among individuals is known as genetic variation. Mutation and genetic recombination are the two major sources of genetic variation. The chapter closely examines these key concepts of genetic variation to provide an extensive understanding of the subject.

Genetic variation can be described as the differences between organisms caused by alternate forms of DNA. Genetic variation in combination with environmental variation causes the total phenotypic variation seen in a population. The phenotypic variation is what is seen by the observer; the height of a plant for instance. The environmental variation is the difference in what each individual experiences.

Examples of Genetic Variation

Genetic Variation between Individuals

Look at the image of mussel shells below. All of these muscles belong to the same species, meaning they can all interbreed with each other. The differences in their patterns represents the total phenotypic variation in the population. Some of the variation comes from genetics, while some comes from the environment.

Two experiments would be required to find the overall genetic variation in the population. In the first, a single mussel would be cloned many times and placed in variable environments. The specimens would be allowed to grow and they would be observed in adulthood. Because their genetics are identical, the variation seen can be attributed solely to environmental variation. In the second experiment, the total variation in a population of wild mussels in the same environment must be observed. At the end of these two experiments, the scientist would have two numbers: One describing the environmental variance and one describing the phenotypic variance.

To get the genetic variation found in this population of mussels, the scientist would simply need to subtract the environmental variance observed with the clone from the total variance observed in the wild population. Another way of calculating the genetic variation is to sample the DNA of the population and measure the differences in the DNA directly. Since genetic variation is produced by differences in the DNA, these differences can be used in reverse to calculate the environmental variation in a population.

Genetic Variation between Species

While the above example discusses genetic variation between members of a population, the concept of genetic variation can be applied on a much grander scale. Consider for instance the Homeobox gene family. This family, known as the "Hox genes" for short, direct and coordinate the positions of body parts during development. These genes, or a variation of them is found among all bilaterally symmetrical animals. This includes everything from insects to fish and mammals. Scientist theorize that an early ancestor developed the Hox genes, which were quickly adapted to many forms of organism. The genetic variation represented in these genes is huge. They produce the different body types of most of the organism on Earth. However, they are still all related and the variance between them can be measured.

Sources of Genetic Variation

With all of the natural variety in the world, it is weird to think that all of the genetic variation comes from only a few simple sources. The simplest source is mutation. As DNA is exposed to the various chemicals and electromagnetic energies of the world, it can mutate. DNA is made of a specific sequence of nucleotides, which produce proteins. Mutations change these proteins by changing the sequence of nucleotides. While mutations are often thought of in a negative context, they drive evolution by putting forth new variants to be tested by the environment.

In fact, genetic variation is so important for species that many species reproduce sexually to aid the process of producing new varieties. Sexually reproducing organisms carry two copies of the genome, allowing mutations to lie dormant or express themselves more subtly. During sexual reproduction, genes are recombined in new ways. This process, known as recombination, shuffles the alleles present and allows different combinations to be expressed. This adds to the total genetic variation. When observing an isolated population, immigration can also be a source of genetic variation. Organism may bring new alleles that have established elsewhere and introduce them to the population.

Human Genetic Variation

Human genetic variation is the hereditary contrasts in and among populations. There may be multiple variations of any given gene within the human population (alleles), a situation called polymorphism. No two people are hereditarily identical. Indeed monozygotic twins (who create from one zygote) have occasional hereditary differences due to transformations occurring during development and gene copy-number variation. Differences between people, indeed closely related individuals, are the key to strategies such as genetic fingerprinting. The study of human genetic variation has developmental significance and therapeutic applications. It can help researchers get it ancient human populace migrations as well as how human groups are naturally related to one another. Modern discoveries appear that each human has an average of 60 new mutations compared to to their parents.

Causes of Variation

Causes of differences between individuals include independent assortment, the exchange of genes (crossing over and recombination) during reproduction (through meiosis) and different mutational events. There are at least three reasons why hereditary variety exists between populations. The natural choice may confer an adaptive advantage to people in a particular environment if an allele provides a competitive advantage. Alleles under selection are likely to occur only in those geographic districts where they confer an advantage. A second important process is genetic drift, which is the impact of irregular changes within the gene pool, under conditions where most mutations are natural (that is, they do not appear to have any positive or negative selective impact on the organism). Finally, little migrant populaces have statistical differences—call the founder effect—from the overall populaces where they originated; when these vagrants settle new zones, their descendant populace typically vary from their population of origin.

Significance of Human Genetic Variation

Nearly all human genetic variation is generally insignificant biologically; that is, it has no adaptive importance. A few variations (for example, a neutral transformation) modify the amino acid sequence of the resulting protein but produce no detectable change in its work. Other variation does not indeed change the amino acid sequence.

Population Genetics

Population genetics is the study of genetic variation within populations, and involves the examination and modelling of changes in the frequencies of genes and alleles in populations over space and time. Many of the genes found within a population will be polymorphic - that is, they will occur in a number of different forms (or alleles). Mathematical models are used to investigate and predict the occurrence of specific alleles or combinations of alleles in populations, based on developments in the molecular understanding of genetics, Mendel's laws of inheritance and modern evolutionary theory.

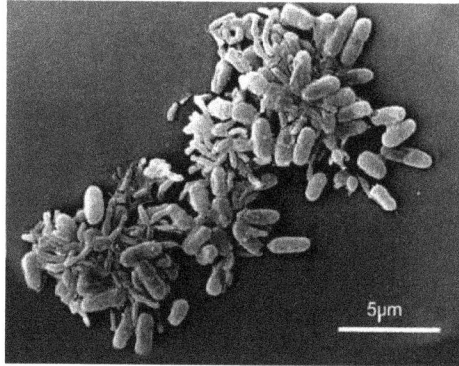

Electron micrograph of a natural bacterial population, showing a range of size and morphology.

The collection of all the alleles of all of the genes found within a freely interbreeding population is known as the gene pool of the population. Each member of the population receives its alleles from other members of the gene pool (its parents) and passes them on to other members of the gene pool (its offspring).

Factors influencing the genetic diversity within a gene pool include population size, mutation, genetic drift, natural selection, environmental diversity, migration and non-random mating patterns. The Hardy-Weinberg model describes and predicts a balanced equilibrium in the frequencies of alleles and genotypes within a freely interbreeding population, assuming a large population size, no mutation, no genetic drift, no natural selection, no gene flow between populations, and random mating patterns.

In natural populations, however, the genetic composition of a population's gene pool may change over time. Mutation is the primary source of new alleles in a gene pool, but the other factors act to increase or decrease the occurrence of alleles. Genetic drift occurs as the result of random fluctuations in the transfer of alleles from one generation to the next, especially in small populations formed, say, as the result adverse environmental conditions (the bottleneck effect) or the geographical separation of a subset of the population (the founder effect). The result of genetic drift tends to be a reduction in the variation within the population, and an increase in the divergence between populations. If two populations of a given species become genetically distinct enough that they can no longer interbreed, they are regarded as new species (a process called speciation).

In many cases, the effects of natural selection on a given allele are directional. The allele either confers a selective advantage, and spreads throughout the gene pool, or it confers a selective disadvantage, and disappears from it. In other cases, however, selection acts to preserve multiple alleles within the gene pool and a balanced equilibrium is observed. This situation, labelled balanced polymorphism, can arise because of a selective advantage for individuals heterozygous for a given allele. For example, the disease sickle cell anaemia is caused by a mutation in one of the genes responsible for the production of haemoglobin. Individuals with two copies of the mutant gene for sickle haemoglobin (HbS/HbS) develop the disease. Individuals that are heterozygous - one copy of the sickle gene and one copy of the normal gene (HbS/HbA) - are carriers of the condition. It is believed that these heterozygous individuals are more resistant to malaria than individuals homozygous for the normal gene (HbA/HbA), and that this selective advantage maintains the presence of the HbS gene in the population. As a result of balanced polymorphism, the gene pools of most populations contain a number of deleterious alleles that reduces the overall fitness of the population (known as the genetic load).

Genetic variation within populations and species can now be analysed at the level of nucleotide sequences in DNA (genome analysis) and the amino acid sequences of proteins (proteome analysis). The genetic differences between species can be used to infer evolutionary history, on the basis that the closest relatives will have gene pools that are most similar.

Genetic Drift

Genetic drift is a process in which allele frequencies within a population change by chance alone as a result of sampling error from generation to generation. Genetic drift is a random process that can lead to large changes in populations over a short period of time. Random drift is caused by recurring small population sizes, severe reductions in population size called "bottlenecks" and founder events where a new population starts from a small number of individuals. Genetic drift leads to fixation of alleles or genotypes in populations. Drift increases the inbreeding coefficient and increases homozygosity as a result of removing alleles. Drift is probably common in populations that undergo regular cycles of extinction and recolonization. This may be especially important in natural ecosystems where both plants and pathogens are likely to have a patchy distribution where each patch is a small population.

Because allele frequencies do not change in any predetermined direction in this process, genetic drift is also called as "random drift" or "random genetic drift."

Measuring Genetic Drift

The magnitude of genetic drift depends on N_e, the effective population size, for the population. N_e is rarely the actual number of individuals in the population (also called

N or the census size). N_e is a theoretical number that represents the number of genetically distinct individuals that contribute gametes to the next generation. N_e can also be thought of as the number of genetically distinct interbreeding individuals in a population. N_e is not easy to quantify because it is affected by reproduction and breeding strategies (inbreeding, outcrossing, asexual reproduction), and is dependent on the geographical area over which a population is sampled.

The genetic drift in a population can be calculated if the effective population size is known. The expected variance in the frequency of an allele (call this frequency p) subject to genetic drift is:

$$\text{Var}(p) = \frac{pq}{2N_e}$$ after one generation of genetic drift for diploid organisms.

After many generations of genetic drift, an equilibrium will be reached. At equilibrium we expect that:

$$\text{Var}(p) = p_0 q_0$$

Where p_0 and q_0 are the initial frequencies of the two alleles at a locus.

If $p_0 = q_0 = 0.5$ and $N_e = 50$ then Var (p) = 0.0025

The standard deviation of (p) = $(0.0025)^{0.5}$ = 0.05.

The standard deviation is the average absolute value of the expected difference among populations after one generation of drift and is approximately equal to the expected change in allele frequency (Δp) within each population. Thus in a population of 50 individuals, with two alleles beginning at equal frequencies, we expect the allele frequencies to change by about 5% each generation.

The degree of change increases as the population size decreases.

If $p_0 = q_0 = 0.5$ and $N_e = 5$ then Var (p) = 0.05

The standard deviation of (p) = $(0.05)^{0.5}$ = 0.22.

In this case, a population that has only five individuals is expected to experience random changes in allele frequencies of about 22% each generation.

Genetic Drift Decreases Gene Diversity and Leads to Population Subdivision

The chance of fixing an allele due to genetic drift depends on the effective population size as well as the frequency distribution of alleles at a locus. To "fix" an allele means that the allele is present at a frequency of 1.0, so all individuals in the population have the same allele at a locus. Large effective population sizes and an even distribution in allele frequencies tend to decrease the probability that an allele will become fixed.

Alleles that occur at a low frequency are usually at a disadvantage in the process of genetic drift. Low-frequency alleles face a higher probability of disappearing from a population than alleles that occur at a higher frequency. Under a scenario of pure genetic drift, the probability of fixation of an allele in a population is its initial frequency in the population. If the initial frequency of an allele is 0.01, then there is a 1% chance that this allele will be fixed in this population. Thinking in a different way, if the initial allele frequency was 0.01 in 100 different populations, then we expect that approximately 1 of those populations would become fixed for this allele after many generations of random genetic drift.

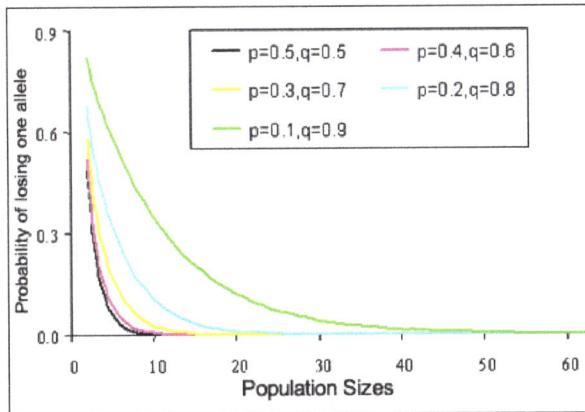

The probability that an allele will drift away in any single generation in a two-allele model with different initial frequencies and different effective population sizes.

The consequences of genetic drift are numerous. It leads to random changes in allele frequencies. Drift causes fixation of alleles through the loss of alleles or genotypes. Drift can lead to the fixation or loss of entire genotypes in clonal (asexual) organisms. Drift leads to an increase in homozygosity for diploid organisms and causes an increase in the inbreeding coefficient. Drift increases the amount of genetic differentiation among populations if no gene flow occurs among them.

Genetic drift also has two significant longer-term evolutionary consequences. Genetic drift can facilitate speciation (creation of a new species) by allowing the accumulation of non-adaptive mutations that can facilitate population subdivision. Drift also facilitates the movement of a population from a lower fitness plateau to a higher fitness plateau according to the shifting balance theory of Sewall Wright.

The amount of population subdivision is expected to increase because of the random losses of alleles that occur in different populations. In addition, random changes in allele frequencies are expected to occur in different populations, and these random changes tend to make populations become differentiated. Finally, small effective population sizes increase the likelihood that mating events will occur between close relatives, leading to an increase in inbreeding and subsequent loss of heterozygosity.

Genetic Load

Genetic load is the reduction in the mean fitness of a population relative to a population composed entirely of individuals having optimal genotypes. Load can be caused by recurrent deleterious mutations, genetic drift, recombination affecting epistatically favourable gene combinations, or other genetic processes. Genetic load potentially can cause the mean fitness of a population to be greatly reduced relative to populations without sources of less fit genotypes. Mutation load can be difficult or impossible to measure. Many species have mutation rates low enough that substantial genetic load is not expected, but for others, such as humans, the mutation rate may be great enough that load can be substantial. In extremely small populations, drift load, caused by the fixation by drift of weakly deleterious mutations, can threaten the probability of persistence of the population. Migration from other populations adapted to different local conditions can bring in locally maladapted alleles, resulting in migration load.

References

- Genetic-variation: biologydictionary.net, Retrieved 6 April, 2019

- Human-genetic-variation: meconferences.com, Retrieved 12 May, 2019

- DNA-repair, science: britannica.com, Retrieved 17 January, 2019

- Population-genetics: le.ac.uk, Retrieved 9 June, 2019

- GeneticDrift: apsnet.org, Retrieved 26 August, 2019

Permissions

Index

www.ingramcontent.com/pod-product-compliance
Lightning Source LLC
Chambersburg PA
CBHW061951190326
41458CB00009B/2842